Fiber Reinforced Composites

1995

)6

6

Fundamental Principles of
FIBER REINFORCED COMPOSITES

K. H. G. Ashbee, B.Sc., Ph.D., D.Sc.

Ivan Racheff Chair of Excellence
University of Tennessee
Knoxville, Tennessee

TECHNOMIC
PUBLISHING CO., INC.

LANCASTER · BASEL

Published in the Western Hemisphere by
Technomic Publishing Company, Inc.
851 New Holland Avenue
Box 3535
Lancaster, Pennsylvania 17604 U.S.A.

Distributed in the Rest of the World by
Technomic Publishing AG

Printed in the United States of America
10 9 8 7 6 5 4 3 2 1

Main entry under title:
 Fundamental Principles of Fiber Reinforced Composites

A Technomic Publishing Company book
Bibliography: p.
Includes index p. 365

Library of Congress Card No. 88-51424
ISBN No. 87762-597-2

Table of contents

Foreword

This is a teaching text for senior undergraduate and graduate engineering students. Each chapter is built around basic physical principles, for which the author goes back to original publications in order to seek out the concepts that have stood the test of time. With fiber composite materials, the engineer has the unique opportunity to design the material as well as the component to be manufactured. This feature of fiber composites is well illustrated by liberal use of actual design studies and case histories. There are also many worked examples and further examples for private study.

Like many teaching texts, this book has been developed from lecture courses given, in this case, at Texas A&M University in 1985, at the Universities of Bristol and Cape Town in 1986, and at the University of Tennessee in 1987.

The author's interest in fiber composites dates from his co-discovery, with F. C. Frank and R. C. Wyatt in the 1960s, of osmosis in composite materials. He has industrial and teaching experience from both sides of the Atlantic, including with Corning Glass Works, the UK Central Electricity Generating Board, Mellon Institute and UC Berkeley.

Readership: mechanical, civil and aeronautical engineering, engineering physics, materials science.

Preface

The purpose of fiber reinforcement is different for each of the classes of matrix materials. For polymers, it is to impart stiffness and strength; for metals, it is to inhibit plastic deformation – particularly creep; and for ceramics, it is to introduce a measure of toughness. One reason for writing the present book is to teach some of the physical mechanisms whereby each of these goals can be achieved.

My second reason for writing this book is to develop the philosophy that, with fiber composites, the engineer has the opportunity to design the material as well as the component or structure. How best to do this requires a thorough understanding of each of the following properties: anisotropy of stress, anisotropy of elasticity, and anisotropy of strength of fiber reinforced materials. In many books on composites, particularly reference books, these properties are dealt with by numerical methods. Here, however, analytical rather than numerical methods are pursued because they better lend themselves to physical interpretation of fundamental concepts. Wherever appropriate, I have used fully worked examples to illustrate this approach.

My third aim with this book is to introduce to the student newly evolving technologies in non-destructive evaluation, in-service monitoring, and failure analysis of composite materials. As with my two other aims, I have approached these topics from the standpoint of the basic underlying physics and have made use of case histories to illustrate applications.

In line with the recommendations of the 1960 Conference Generale des Poids et Mesures, the body responsible for maintaining standards of measurements, now adopted worldwide by the engineering community, I have used the SI system of units. SI, which is the abbreviation in many languages for Systeme International d'Unites, is an extension and refinement of the traditional metric system. It embodies features which make it logically superior to any other system as well as practically more convenient. The main features of SI are as follows:

1. There are six basic units:

Physical quantity	Name of unit	Symbol for unit
length	metre	m

mass	kilogram	kg
time	second	s
electric current	ampere	A
thermodynamic temperature	kelvin	K*
luminous intensity	candela	cd

2. The unit of force, the newton (N $=$ kg m s^{-2}), is independent of the Earth's gravitation and the often confusing introduction of **g** into equations is no longer necessary. This has enormous advantages in the teaching of composite materials; **g** is not constant over the surface of the Earth and, in aerospace applications of composite materials, can take a range of values from several times its Earth value at launch to near-zero when deployed in space.
3. The unit of energy in all forms is the joule (N m), and of power is the joule per second (watt).
4. "Electrostatic" and "electromagnetic" units are replaced by SI electrical units.
5. Multiples of units are restricted to steps of a thousand, and fractions to steps of a thousandth:

Multiplying factor	Prefix	Symbol
10^{12}	tera	T
10^{9}	giga	G
10^{6}	mega	M
10^{3}	kilo	k
10^{-3}	milli	m
10^{-6}	micro	μ
10^{-9}	nano	n
10^{-12}	pico	p
10^{-15}	femto	f
10^{-18}	atto	a

The publication explosion in composite materials makes it impractical to direct the reader to a small number of key papers for advanced study; most contemporary authors have published many papers. I have, therefore, opted to put the onus on the reader to select his topics of interest and then make use of computer searches to identify specific papers.

Much of the groundwork for this book has its origins in the physics of materials Master's program that ran at the University of Bristol in the 1960s and 1970s. I have drawn heavily on the teachings of my Bristol colleagues, particularly F. C. Frank and J. F. Nye. When I was an undergraduate at Birmingham University, my tutor, the late J. D. Eshelby, used to say that the wrong people write books. I suspect that both Frank and Nye could do superior jobs on the

*By international agreement, the degree symbol is omitted when temperature is specified in kelvins.

subject of composite materials, and I take full responsibility for any gaps in my recounting of their lectures and discussion remarks. When developing the lecture courses on which the book is based, I also re-cast, in the context of composite materials, many problems set by colleagues at Bristol as examination questions and tutorial exercises, for which I am indebted to M. V. Berry, D. F. Gibbs, J. C. Gill and M. G. Priestley. I would also like to thank the UCLA short course program instructors in composite materials, particularly P. W. R. Beaumont, J. C. Halpin, and K. T. Kedward for sustaining my interest in the teaching of composite materials over the past two decades. The line drawings were stencilled by D. A. Tossell, the manuscript was word-processed by E. Walter, and draft chapters were read and commented on by J. R. Brewster, S. M. Joslin, D. A. Tossell, E. Walter, R. C. Wetherhold, and Z. R. Xu. I gratefully acknowledge the time each of these people contributed to the preparation of this book.

Ken Ashbee
UT Knoxville
December 1987

1

Specific strength and specific modulus

"...the unit for stress ought not to contain a concealed g ..."[1]

The words *glassfiber reinforced plastic* and *resin-bonded fiberglass* mean what they say. They describe the same material, but, whereas the former reminds us that the glass contributes much of the strength and particularly the modulus, the latter alludes to the fact that the plastic counters the brittleness of the glass. The one sees the fibers as the means for bolstering the strength and stiffness of the plastic, whereas the other sees the plastic as adhesive by which the fibers are bundled together but separated from one another in order to exploit the fact that the misfortune to find a crack in any one fiber is of little consequence since that crack has to stop at that fiber surface; in a monolithic block of glass, a crack is likely to propagate all the way to the free surface.

When designing composite materials with which to manufacture structures of minimum weight, airplanes for example, the ultimate tensile strength (*UTS* = maximum load sustained before failure/original cross-sectional area) and tensile stiffness (*E*) for a given weight are more relevant than are the absolute magnitudes of these properties. The Young's modulus to density ratio

$$\frac{E(\text{N m}^{-2})}{\varrho(\text{kg m}^{-3})} \tag{1.1}$$

has dimensions of (velocity)^2 (m^2 s^{-2}) cf.:

$$V_P = \sqrt{\frac{E}{\varrho}} \tag{1.2}$$

which is the velocity of longitudinal sound along the fiber. Sound propagates through solids by way of elastic vibrations of the atoms and, when these vibra-

---

[1] F. C. Frank, "The Strength of Polymers," *Proc. Roy. Soc.*, A 282:10–16 (1964).

1

tions are in the direction of propagation, they are said to be longitudinal vibrations. E/ϱ is known as the specific modulus.* Following F. C. Frank,[1] since $\sqrt{UTS/\varrho}$ is a velocity but not of any known wave, the strength to weight ratio has more physical significance if expressed by

$$\frac{UTS(\text{N m}^{-2})}{\varrho(\text{kg m}^{-3})g(\text{m s}^{-2})} = L_{max} \qquad (1.3)$$

which has dimensions of length (meters), and is the longest length of fiber of uniform cross-section that could be suspended in the Earth's gravitational field if there were no wind. L_{max} is known as the specific strength. This definition assumes that "g" is constant.

Strength to weight ratio and specific modulus data for some common fibers are shown in Figure 1.1.

It might be concluded from consideration of the specific strengths of the common engineering materials that any structure whose main function is to support its own weight, a bridge for example, should be made from nylon, nylon being a relatively inexpensive material ($L_{max} \cong 70$ km for nylon cf. $L_{max} \cong 10$–20 km for steel). However, it turns out that nylon is not sufficiently stiff to withstand as much wind pressure as is steel. Polymeric materials do not, in general, have very high elastic modulus; they are strong only in the sense that they can be loaded to high elastic stress before breaking. A nylon compression member bends unless extra stiffness has been designed into it by its geometry, for example, by making it in the form of a tube. The designing in of stiffness by using tubes instead of rods is a well practiced design feature of, for example, bridges; whereas a rod of steel might bear the design load, it is more liable to buckle and bend than is a steel tube. This resistance to buckling advantage of tubes is widely exploited in fiber reinforced laminate technology both in the form of filament wound tubes and in the construction of honeycomb core sandwich structures. Thus, the steel rods in the Brooklyn Bridge could be replaced by equivalent nylon tubes but these tubes would have outside diameters about ten times that of the steel rods, and would incur much higher wind drag.

J. E. Gordon[2] points out that many materials have about the same modulus/weight ratio. Only carbon, asbestos and aramid fibers offer substantial advantages over other materials.

---

*A less than rigorous conception perpetrated in the literature is that specific modulus and specific strength have dimensions lb in^{-2}. This is unsatisfactory on two counts. First, the pound (lb) is not a unit of force. It is a unit of mass and calling it a unit of force not only contains a concealed "g" but also assumes that "g" is the same on all parts of the Earth. Second, the implication that density is a dimensionless quantity requires that it be defined relative to the density of water and incurs a second approximation, namely that the density of water is independent of temperature.

[2]J. E. Gordon. *The New Science of Strong Materials*, Penguin, pp. 234–254 (1968).

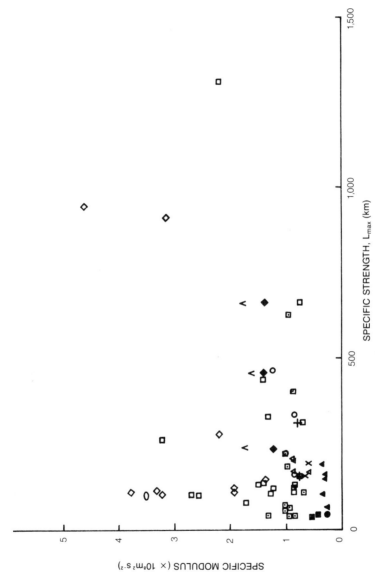

Figure 1.1. Specific modulus (velocity²) versus specific strength (L_{max}) for several fibers. For comparison, data points for some more conventional materials are also included. □ – SiC + – silicon ○ – Si_3N_4 △ – Kevlar, X – nextel, ■ – fiberfrax, ▲ – glass, ◇ – carbon, ◆ – BeO, ∧ – boron, □ – AlN, ○ – alumina, ○ – beryllium, ● – iron. Figure compiled by S. M. Joslin.

1.1 Physical origins of strength

A useful starting point for comparing the theoretical strengths of materials is the energy of bonding which, for the elemental solids at least, is proportional to the temperature required to break the bond between atoms. The lighter rare gases, and hydrogen, boil at below 100 K whereas the strongly bonded transition metals boil at 3,000–4,000 K (to be precise helium and hydrogen boil at 4.2 and 20 K respectively and, at the other extreme, tungsten boils at 5693 K) from which we deduce that there is in nature a range of some 30 to 40 times in the energy of interatomic bonding, so we might expect a similar range of 30 to 40 times in the modulus and strength of solids.

The forces between electrostatically neutral molecules, the van der Waals forces, have their origins in the so-called overlap forces, London[3] forces, and Debye[4] forces. To get a feel for the nature of these three forces, and hence for the nature of bonding between atoms and molecules in general, it is instructive to consider interatomic bonding in the simplest of all solids, solid argon.

Overlap forces are repulsive forces, which vary as r^{-12} where r is the interatomic distance, and give molecules a finite size. When we push negative electrons into the overlap space, we have to do work against the Coulomb repulsion or else the kinetic energy of the electrons increases.

Debye forces are attractive forces. Some molecules are electrostatically dipoles—HCl for example. For these we can calculate statistically the degree of preferred orientation in an electric field. The dipole moment can be measured from the temperature dependence of the dielectric constant. Re-stated in quantum mechanics terminology, we have split energy levels for rotating molecules. Because it is dipolar, HCl has an electrostatic field around it. HCl molecules are therefore likely to take up special orientations relative to each other. This orientation amounts to a fairly long-range attractive force between molecules.

To understand the origin of dipoles, consider the water molecule. To avoid electrostatic repulsion, the two protons go to opposite sides of a diameter, see Figure 1.2. If we rotate the molecule, the electrons are dragged to one side as indicated by the arrows in Figure 1.1 which gives rise to the polarity indicated by the + − symbols. This polarity is equivalent to inducing a dipole moment on the oxygen atom. Therefore the water molecule is polar; it has a dipole moment, and should be denoted:

Debye forces are, then, attractive forces arising from electrostatic forces, typically from dipoles. Returning to argon, there is no possibility of a dipole, so how are argon atoms bonded together in solid argon? We could invoke quadrupoles or

[3]F. London. "On Centers of van der Waals Attraction," *J. Phys. Chem.*, 46:305–316 (1942).
[4]P. Debye. "Die van der Waalsschen Kohasionskrafte," *Physik Zeitschr.*, 21:178–187 (1920).

p

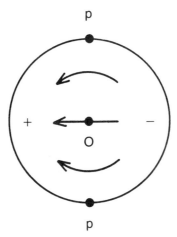

p

Figure 1.2. Illustrating the forces between an oxygen atom and two protons.

octopoles, but these are not stable. London forces is the answer. It follows from Heisenberg's uncertainty principal that the center of gravity of the "electron cloud" does not coincide with the center of gravity of the argon nucleus. As a consequence, we have fluctuating electrostatic moments and therefore a fluctuating dipole moment. And so too do neighboring atoms, due to induced dipole moments. Hence, they attract. What about a molecule like carbon disulfide? CS_2 —SCS—and there is no dipole moment because of its symmetry. There is however, a fairly strong quadrupole moment. The first term in the London forces series is related to the fluctuating dipole moment on one atom inducing fluctuating dipole moments on a neighboring atom and leads to the conclusion that the potential energy between the two is proportional to r^{-6}. If we go to higher terms, which describe fluctuating quadrupole moments, we get a series of potential functions such as proportional to r^{-8}.

So the bonding in rare gas solids is due to London forces only, hence the low boiling temperature. And, in general, it is usual to say that the energy arising from van der Waals forces is very small and can be neglected. But this is not really true. At the other extreme, the bonding in tungsten is metallic bonding, hence the high boiling point. As a matter of experimental fact, the observed strength of argon is closer to its theoretical strength than is the observed strength of tungsten.

Thus far we have introduced only van der Waals bonding, and we will return to it in Chapter 11, Section 11.4, when discussing the physics of adhesion between dissimilar materials. Covalent, ionic and metallic bonding also exist in composite materials. To a zero order approximation, all interatomic bonding has its origin in Coulomb attraction; it is the ways in which Coulomb attraction operates that distinguish van der Waals from covalent from ionic from metallic bonding. In principle we could solve, say, the Coulomb attraction for the iron atom, one nucleus plus 26 electrons. However, the calculation is too big to undertake, so crude approximations are made.

A characteristic feature of covalent bonding is that a pair of electrons is shared by neighboring atoms. This is always the case for C−C bonds. However, we do not always have to describe covalent bonding this way. Consider diamond. Figure 1.3 shows the crystal structure of diamond. This feature of each carbon atom bonded to four carbon atoms also occurs in organic chemistry. So, when discussing interatomic bonding of polymers, it is instructive to recall the theory of semiconductors; diamond has the same crystal structure as the elemental semiconductors silicon and germanium. To a zero order approximation then, we consider an array of carbon cores. Each core contains four tightly bound electrons. Add two electrons to one core. The core is tetravalent and, when we add a third electron, the Pauli exclusion principle forbids the same wave function.

(a)   (b)

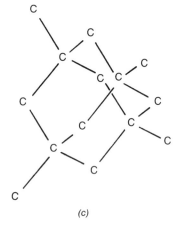

(c)

Figure 1.3. (a) Regular tetrahedron, (b) join the body center of a regular tetrahedron to each of its four corners, (c) the diamond structure is generated by the tetrahedral bonds defined in (b).

Making use of the Hartree approximation leads to the same conclusion about diamond as about silicon and germanium, that is, that diamond is a semi-conductor, albeit with a large band gap.

The same kind of calculation can be done for ionic crystals such as NaCl, but not for MgO; the O^{2-} anion is not stable—there is some covalent bonding. The mean interatomic distances calculated for NaCl are different from those for KCl, and the difference is a measure of the difference in ionic size of potassium over sodium. To account for the absolute values for interatomic distances, use is made of Pauling's[5] table of ionic sizes, although absolute truth should not be attached to Pauling's ionic radii. Slater's[6] table of ionic radii lists magnesium as larger than oxygen, for example, and predicts better lattice parameters for some crystals.

Metallic bonding arises from Coulomb attraction between free electrons and positively charged cores. If the number of free electrons per primitive unit cell is odd, the crystal is a conductor. If the number is even, the crystal can be an insulator.

It turns out that the nature of the bonding is not all that important in determining strength. After all, the stretching of bonds by thermal agitation is enormous compared with the stretching of bonds by applied stress. The arrangement of atoms is much more important. We find ductile behavior, and therefore low strength, in metals that have simple crystal structures. There are an infinite number of close packing arrangements of spheres. The simplest are face centered cubic (fcc) for which there are four atoms per primitive unit cell, and hexagonal close packed (hcp) with two atoms per primitive unit cell. Fcc metals are ductile. Hcp metals are ductile in single crystal form but, since they have only one plane of wavy slip, are brittle in polygranular form because the deforming grains of different orientation jam and block each other. Body centered cubic (bcc) is close packed in the sense of covering power—cf. uniformly distributed currants in a plum pudding. If we keep the kinetic energy of the free electrons high without keeping their potential energy down, we end up with the ion cores in a bcc array. The density is low and the metal, such as sodium and lithium, is ductile. The dense bcc metals, of which iron and tungsten are examples, have bcc structures because the ion cores are close together and give rise to complexity in core-core interaction and magnetic interactions. These bcc metals are not ductile. Details of the reasons for the low ductility of some bcc metals include trigonal splitting of screw dislocations with $\frac{1}{2} < 111 >$ Burgers vectors, and Cottrell pinning of dislocations by impurity atoms. The sodium chloride structure can be described as sodium atoms in the octahedral interstices of fcc chlorine. One example of a crystal with the NaCl structure is AgCl, which is transparent and plastic. AgCl, which deforms plastically by slip on $\{110\}$ wavy planes is about as plastic as aluminum which deforms by slip on $< 110 > \{111\}$ systems. The plastic deformation

---

[5] L. Pauling. "The Sizes of Ions and the Structure of Ionic Crystals," *J. Amer. Chem. Soc.*, 49:765–790 (1927).
[6] J. C. Slater. *Quantum Theory of Atomic Structures, Volume I*, New York:McGraw Hill, p. 210 (1960).

of ZnS illustrates one reason why materials with complicated crystal structures are brittle, in that the zinc atoms have to pass through high energy environments in order to be displaced.

For many applications, materials selection is based on toughness, a property which has to do with the importance of stress concentrations. The physical meaning and importance of toughness will be postponed until Chapter 8.

1.2 Fiber diameters

The fibers, and whiskers, used to reinforce plastics and other materials have diameters of the order of 10 μm. To see why this is the optimum fiber diameter for glass fiber, it is necessary to recount the observations of Griffith.[7] Most solids fracture at applied stresses of the order of 1 kbar (100 MN m^{-2}) and, for materials that are wholly elastic up to the point of fracture, this corresponds to a fracture strain of 0.01%. Notable exceptions are the naturally occurring composite materials wood and bone, both of which can exhibit fracture strains of up to 1%. Now the theoretical strength of solids is of the order of 100 kbars (10 GN m^{-2}) which, for an elastic body, corresponds to a fracture strain perhaps as large as 20%. So solids are inherently weak and, somewhat unexpectedly, this weakness depends on the dimensions of the body. For example A. A. Griffith[7] found that the tensile strength of glass test-pieces varies markedly with test-piece diameter, see Figure 1.4.

The high strength of very small diameter test-pieces (fibers) is understandable since, in the limit, a single unbroken chain of atoms must have the theoretical tensile strength. The dilemma that occurred to Griffith is "why is thick glass weak?" The answer, also proposed by Griffith, is that glass, and solids in general, contain built-in stress raisers.

Fiberglass is drawn vertically downward through orifices from molten glass contained in platinum/rhodium tanks. As they emerge from the orifices, the fibers are sprayed with water to dissipate static electricity, and with (usually) an aqueous solution of silane coupling agent (size). When wound as a bundle onto a bobbin, the fibers are collectively referred to as a roving.

Carbon fiber in its continuous fiber form is typically 7 μm in diameter. There are three fabrication routes for carbon fiber, the respective starting materials for which are (i) rayon (cellulose), (ii) polyacrilonitrile (PAN), and (iii) pitch. The latter can be either isotropic if spun from the melt or anisotropic, which imparts high modulus, if spun from the liquid crystal mesophase. There are five stages in the manufacture of all three fibers, namely (a) spinning, which may be performed wet or dry, or from the melt; (b) drawing, which is done above room tempera-

[7]A. A. Griffith. "The Phenomena of Rupture and Flow in Solids," *Phil. Trans. Roy. Soc.*, A 221:163–198 (1921).

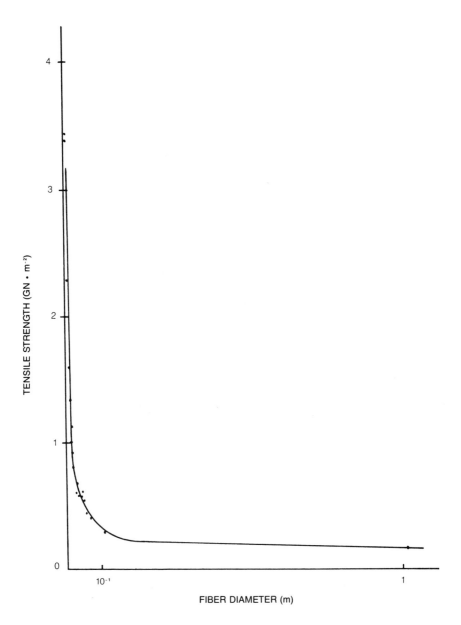

Figure 1.4. Strength of glass fibers. Data points taken from Table V of A. A. Griffith.[7]

ture, usually at between 100°C and 300°C. W. Watt[8] was the first to document the effect of fiber extension on crystalline orientation and its effectiveness in controlling modulus of the final fiber; (c) stabilization by way of oxidation by heating to 400°C. The 400°C oxidation stage is A. Shindo's[8] lasting contribution to the technology. It significantly lowers the overall weight loss and thereby ensures higher graphite yield and better properties; (d) carbonization at temperatures in the range 1000–2000°C; (e) graphitization at 2000–3000°C.

An untwisted bundle of carbon fiber is usually referred to as a tow, a term borrowed from textile technology. In practice, tows are often deliberately twisted in order to improve the ease of handling.

The mechanical properties of the fibers produced by the three fabrication routes are different, and the differences are summarized in Figure 1.5. The main disadvantage offered by carbon fiber manufactured from pitch is poor axial compression strength. This is attributed to buckling following changes in the relatively large single crystals characteristic of pitch-based fibers. On the other hand, pitch-based fiber exhibits remarkably high electrical conductivity, of order 1.5 kW m^{-1}K^{-1} in the axial direction which is about four times larger than the thermal conductivity of copper ($\sim$ 11 times on a weight for weight basis).

In addition to continuous fiber, carbon fiber can be prepared in the form of graphite whiskers, the first invention of which is generally attributed to Bacon.[8] These are single crystal basal slabs of graphite, the morphology of which is not unlike that of a partially unravelled scroll. Graphite whiskers have an axial modulus of $\cong 690$ GN m^{-2} and an axial strength of $\cong 20$ GN m^{-2}.

Firing of the precursor involves an oxygen pick-up of $\cong 8$ wt% and results in a 50% decrease in the fiber diameter. Attempts have been made to make the final diameter closer to the 10 μm optimum established for glassfiber. However, the mechanism for conversion to graphite is diffusion controlled and, whereas conversion of 15 μm diameter PAN filaments to 7 μm diameter carbon fiber might take four hours, conversion of 30 μm diameter filaments to 15 μm diameter fibers is found to take about forty hours. In addition, an unacceptably higher proportion of the larger diameter fibers takes the form of hollow tubes rather than solid rods.

The 10 μm order of magnitude for the diameter of high modulus polymer fibers ($\cong 12.5$ μm for Kevlar 49®*) has its origin in the drawing and fiber spinning processes by which highly oriented polymer fibers are manufactured. Aramid fibers, such as Kevlar, are composed of chains of aromatic carbon rings linked by $-CO-NH-$ groups. To understand why highly oriented polyethylene, for example, is endowed with high modulus, consider diamond. The Young's modulus for diamond in the [110] direction is 11.6 Mbar (1.16 TN m^{-2}). Figure 1.6 shows a model of the diamond structure, viewed along [110]; the structure is composed of fully aligned zig-zag chains of carbon atoms just like those in polyethylene, Figure 1.7.

---

[8]Carbon fibers were independently invented at about the same time by A. Shindo in Japan, by R. Bacon in the USA and by W. Watt in England, although Thomas Edison had experimented with carbonized cotton filaments for incandescent lamps before the turn of the century.

*Kevlar is a registered trademark of E. I. du Pont de Nemours.

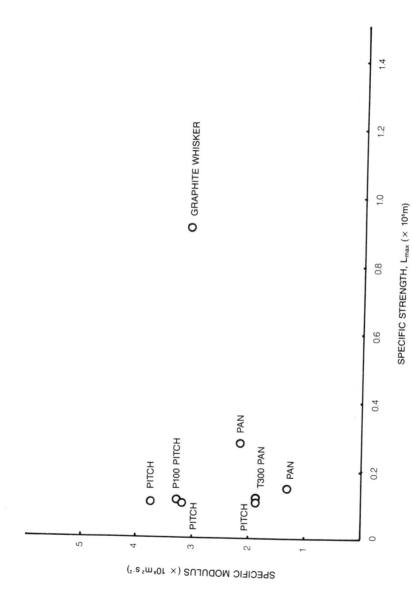

Figure 1.5. Specific modulus versus specific strength for different carbon fibers. Data compiled by S. M. Joslin.

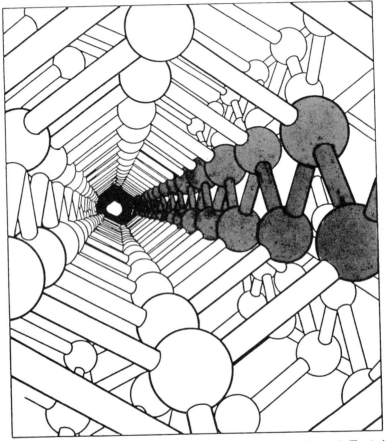

Figure 1.6. Diamond structure viewed along [110]. After L. Pauling and R. Hayward, *The Architecture of Molecules*, W. H. Freeman, San Francisco (1964).

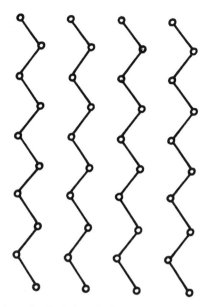

Figure 1.7. Crystal structure of polyethylene $(-CH_2-CH_2-)_n$. General view showing zig-zag chains of carbon atoms. After C. W. Bunn, "The Crystal Structure of Long-Chain Normal Paraffin Hydrocarbons. The 'shape' of the $>CH_2$ group," *Trans. Faraday Soc.*, 35:428–491 (1939).

In diamond, only two of the four tetrahedrally oriented first nearest neighbor bonds are utilized by the [110] zig-zag chain; the other two are at right angles to [110] and contribute nothing to the Young's modulus in this direction. By the same token, the carbon-to-hydrogen bonds in fully aligned polyethylene contribute nothing to its longitudinal Young's modulus. The cross-sectional area per chain in diamond is 0.0448 (nm)^2 or 4.48 Å^2, four times smaller than that in polyethylene,

Table 1.1 Classification of silicates.

Oxygen:Silicon	Structural Unit	Silicate Class
4:1	$(SiO_4)^{-4}$	nesosilicates
3.5:1	$(Si_2O_7)^{-6}$	sorosilicates
3:1	$[(SiO_3)^{-2}]_n$	cyclosilicates (n-member rings) inosilicates (n-member chains)
2.5:1	$(Si_4O_{10})^{-4}$	phyllosilicates (layers)
2:1	$[SiO_2]_n$	tektosilicates (3-dimensional skeletal cages)

Figure 1.8. Two-dimensional chain structure of corner-sharing $[SiO_4]^{4-}$ tetrahedra.

0.182 $(nm)^2$ or $18.2 \, Å^2$. Hence, from this analogy, a modulus of 2.85 Mbar (285 GN m^{-2}) might be expected for fully aligned polyethylene. This compares with 0.69 Mbar (69 GN m^{-2}) for highly oriented polyethylene[9] and 1.3 Mbar (130 GN m^{-2}) for Kevlar 49®.

It was pointed out at the beginning of this chapter that asbestos, carbon and aramid fibers stand above all other materials in the specific modulus league table. Asbestos is a naturally occurring silicate fiber and, to understand the origin of its $\cong 10 \, \mu m$ diameter, it is necessary to recall some silicate chemistry. One classification of silicates is in descending order of oxygen/silicon atom ratio (see Table 1.1). The basic building block is the $(SiO_4)^{4-}$ tetrahedron and, at oxygen:silicon ratios below 4:1, adjacent tetrahedra share corners to form structural units characteristic of the silicate class. There are three commonly occurring forms of asbestos. Tremolite, $Ca_2Mg_5Si_8O_{22}(OH)_2$, is an inosilicate with crystal structure based on the double chain of $(SiO_4)^{-4}$ tetrahedra shown in Figure 1.8. Notice that the apexes are all pointing upwards. The fiber diameter is determined by the double chain width. Crocidolite, $Na_2Fe_3[II]Fe_2[III]Si_8O_{22}(OH)_2$, the blue-colored form of asbestos well-known as the cause of the medical condition asbestosis, has a somewhat similar structure. Chrysotile, $Mg_3Si_2O_5(OH)_4$, is structurally different from the other two forms of asbestos. It is a phyllosilicate in which the layers wrap around on themselves to form capillary tubes typically several centimeters long and $< 10 \, \mu m$ internal diameter. The capillary diameters are 500 Å for the magnesium silicate form of chrysotile. Substitution of cobalt and nickel for magnesium gives rise to capillary diameters of 360 Å and 700 Å, respectively.

1.3 Weight for weight comparison of structural materials

In Figure 1.1, data points for the fibers commonly used to manufacture composite materials are well distributed in the high strength and high modulus regions of the plot. Composite materials manufactured from these fibers are nearer the origin and therefore closer in properties to the more conventional materials including the metals. To see why this is the case for modulus, consider in Figure 1.9, the packing fraction (η) for a bundle of fibers (identical parallel cylinders).

If R is the fiber radius, the area of the "unit cell" outlined by the parallelogram is $R\sqrt{3} \times 2R = 2\sqrt{3}R^2$. Since the unit cell contains one fiber,

$$\eta = \frac{\pi R^2 l}{2\sqrt{3} R^2 l}$$

where l is the fiber length.

[9]Molecular orientation can be induced in polyethylene by cold drawing, by extrusion, and by spinning from solution. The first reported modulus measurements were by J. M. Andrews and I. M. Ward, "The Cold-Drawing of High Density Polyethylene," *J. Materials Sci.*, 5:411–417 (1970).

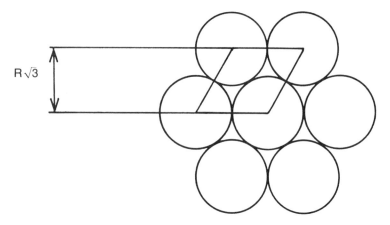

Figure 1.9.

Hence

$$\eta = \frac{\pi}{2\sqrt{3}} \tag{1.4}$$

$$= 0.9069$$

So even in an ideally close-packed uniaxial bundle of fibers bonded together by resin, the fiber loading is only 90.69% of the overall volume. Assuming mechanical continuity at the fiber/matrix interface, the axial displacement and hence the axial strain (ϵ_l) in an axially loaded uniaxial fiber reinforced composite will be identical for both fiber and matrix. However, since fiber and matrix materials have different Young's modulus, the axial stresses will be

$$\sigma_f = E_f \epsilon_l \qquad \text{and} \qquad \sigma_m = E_m \epsilon_l$$

where the subscripts f and m denote fiber and matrix respectively. The overall stress applied to the composite is

$$\sigma_l = \frac{P}{A}$$

where P is the axial load and A is the cross-sectional area of the composite.
Since

$$P = \sigma_f A_f + \sigma_m A_m$$

$$\sigma_l A = E_f \epsilon_l A_f + E_m \epsilon_l A_m \tag{1.5}$$

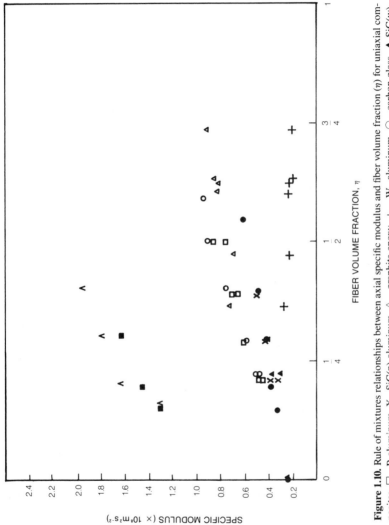

Figure 1.10. Rule of mixtures relationships between axial specific modulus and fiber volume fraction (η) for uniaxial composites. □ — B-aluminum, X — SiC(p)-aluminum, △ — graphite-epoxy, + — W — aluminum, ○ — carbon-glass, ▲ — SiC(w)-aluminum, ● — SiC-tungsten, ■ — graphite-aluminum, ∧ — graphite-magnesium. Figure compiled by S. M. Joslin.

FIBER VOLUME FRACTION, η

SPECIFIC MODULUS ($\times 10^8 \text{m}^2\text{s}^{-2}$)

and the fiber volume fraction and matrix volume fraction are respectively

$$V_f = \eta = \frac{A_f}{A} \; ; \quad V_m = \frac{A_m}{A}$$

Substituting for σ_1, A_f and A_m in Equation (1.5) we get

$$E_1 = E_f V_f + E_m V_m$$

or, since $V_f + V_m = 1$, and $V_f = \eta$ in Equation (1.4)

$$E_1 = E_f \eta + E_m(1 - \eta) \tag{1.6}$$

the so-called rule of mixtures Young's modulus for uniaxial composites. In fiber reinforced plastics, E_f/E_m is in the range 10–100 so, for $\eta = \pi/(2\sqrt{3})$, $E \cong$ 0.9 E_f plus at most 0.01 E_f.

The upshot is that the highest possible specific modulus values for uniaxial fiber reinforced composites are $\cong 91\%$ of the fiber specific modulus values. Figure 1.10 shows modulus versus fiber volume fraction (fiber loading) data for several composite systems.

Note in Figure 1.10 the substantial advantage offered by light-weight metal matrix composites over the more conventional resin matrix composites.

The dilution in ultimate tensile strength with decreasing fiber loading is non-linear (see Chapter 11, Section 11.7), and is usually subject to a measure of variability. Hence the highest possible strength/weight ratios for uniaxial composites are rather less than 90% of the values shown in Figure 1.1 for the respective fibers.

In conventional composites, much lower specific modulus and strength/weight ratios are realized in practice. This is because conventional methods of fabrication do not yield close packing of fibers. With strict quality control, packing fractions close to $\eta = 4/5$ can be achieved with filament winding. However, in laminates fabricated from prepreg, $\eta = 3/5$ is closer to the accepted norm.

Also shown in Figure 1.1 are specific modulus versus strength/weight data for a few metals. It should be noted that the data for quasi-isotropic fiber reinforced laminates overlap those of the new light-weight aluminum/lithium alloys. To fully realize the potential of fiber reinforced composites, it is essential to move away from the quasi-isotropic lay-ups and towards uniaxial composites with high fiber loadings. Note also in Figure 1.1 the very high specific modulus of beryllium.

1.4 Design considerations

When looking for applications for high specific modulus materials, it should be borne in mind that displacement, and hence stresses and strains, generally decrease linearly with increasing E/ϱ and that vibration frequencies and speeds

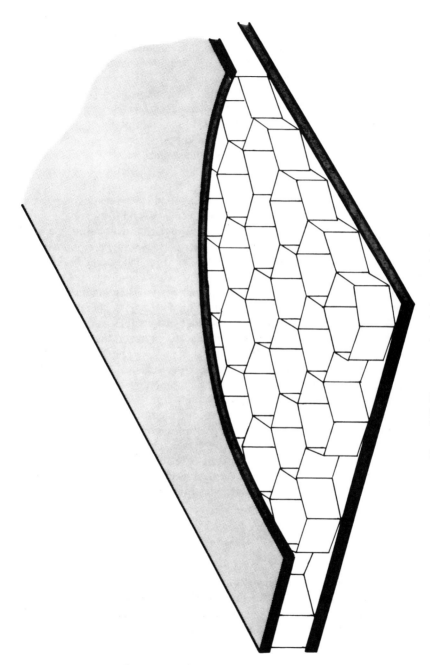

Figure 1.11. The construction of a honeycomb laminate.

of rotation generally go as $\sqrt{E/\varrho}$. A word of warning, though; laminates are usually reinforced in only two dimensions, there is no fiber reinforcement between adjacent plies. Consequently, the modulus, and hence the specific modulus, in the through thickness direction is matrix dominated and may be relatively very low. The modulus in compression is also low. For example, for aramid fiber/epoxy laminates, the 0° Young's modulus in compression is typically 40 GN m^{-2} compared with 70 GN m^{-2} in tension. This fact has important consequences for submarine hulls. When surfaced, a submarine floats because her weight is less than the weight of water that would be displaced if she were totally submerged. When a submarine dives, the ballast tanks have to be filled with water until her total weight equals the weight of her submerged displacement. However, the deeper she dives the higher the water pressure, and because the air pressure inside is maintained at about an atmosphere, the overpressure causes the hull to shrink. That is, the volume of the hull decreases, and if the weight of the submarine plus ballast remains unchanged the submarine will sink deeper and deeper until it is crushed by the increasing water pressure. In conventional steel hull submarines, this implosion is prevented by blowing ballast and working the hydrofoils. However, with fiber reinforced hulls the low compressional modulus means substantially larger volume shrinkage and hence faster sinking.

Graphite fiber is expensive ($44.00/kg in 1986). The onus is, therefore, on the user's materials selection committee to consider all possible ways of avoiding graphite, at least at the preliminary materials selection stage. If the component or structure to be manufactured is stiffness critical, it may well be that, by changing the geometry, the specific stiffness can be raised sufficiently to meet specification. For example, if bending stiffness is required, a sandwich construction consisting of an aluminum or Nomex® honeycomb core bonded between glass-fiber/epoxy resin laminates may be adequate, see Figure 1.11.

There are, of course, other reasons why glassfiber reinforcement may be unacceptable. Glass is prone to static fatigue (stress corrosion), particularly in aqueous environments. Glass is also an insulator, electrical and thermal. In addition, there are differences in the time dependent mechanical properties (damping, creep, fatigue, impact resistance) between glassfiber and graphite fiber reinforced composites.

1.5 Worked examples

1. Air bubbles and vapor-filled voids effectively reduce the resin's contribution to overall modulus in fiber reinforced resins. Use J. D. Eshelby's[10] equations for the elastic field of an ellipsoidal inclusion to estimate the effect of spherical cavities on both the bulk modulus and the shear modulus.

[10]J. D. Eshelby. "The Determination of the Elastic Field of an Ellipsoidal Inclusion, and Related Problems," *Proc. Roy. Soc.*, A 241:376–396 (1957).

Substituting $\sigma = 1/3$ for Poisson's ratio in the equations for a spherical inclusion given at the beginning of section 5 of Eshelby's paper, we get $\alpha = 2/3$, $\beta \cong 1/2$ and, since $\varkappa_1 = \mu_1 = 0$ for a void,

$$A = -\frac{1}{\alpha - 1} = 3$$

$$B = -\frac{1}{\beta - 1} = 2$$

Hence Eshelby's equations for the effective bulk modulus and effective shear modulus become

$$\varkappa_{effective} = \varkappa(1 - 3V)$$

$$\mu_{effective} = \mu(1 - 2V)$$

where V is the volume fraction of spherical cavities.

2. Why is resin bonded Kevlar fiber a popular choice of material for flywheels? Show that, to a first approximation, it makes little difference whether the Kevlar fiber is used in the rim or in the spokes of a flywheel (University of Tennessee Examination, 1987).

First part: Energy density ($J/m^3 = N\ m/m^3 = N/m^2$) has the same dimensions as Young's modulus and UTS, and Kevlar fiber is stiffer (higher Young's modulus) and stronger than any other man-made fiber.

Second part: For the case of annular reinforcement, ignore the weight of the spokes, i.e., reduce the problem to calculation of the circumferential stress in a rotating ring of density ϱ and circumferential speed v.

Consider, in Figure 1.12, an angular sector ϕ of the ring.

The tension on it from the rest of the ring must satisfy the equation

$$\text{inward force, } 2T \sin\left(\frac{\phi}{2}\right) = \text{mass} \times \text{inward acceleration, } \frac{m\ v^2}{r}$$

For small ϕ, this gives

$$T\phi = \varrho A r\ \phi\ \frac{v^2}{r}$$

Therefore stress

$$\sigma = \frac{T}{A} = \varrho\ v^2$$

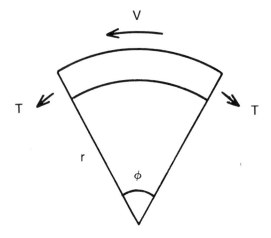

Figure 1.12.

For the case of radial reinforcement, ignore the weight of the rim, that is reduce the calculation to evaluation of the radial stress in a rotating circular disc of density ϱ and circumferential speed v. (This latter part of the calculation is the basis of the so-called brush flywheel.)

The centrifugal force on each radial element is

$$dT = (\varrho A \, dr)\, \frac{r^2 \omega^2}{r}$$

$$d\sigma = \varrho r \omega^2 \, dr$$

$$\therefore \sigma = \frac{\varrho r^2 \omega^2}{2} = \frac{\varrho v^2}{2}$$

From the practical standpoint, the fiber density in the hub of a radially reinforced flywheel might be inconveniently high.

3. Show that the limit to the height of a uniform vertical rod clamped at its lower end is proportional to the cube root of its specific modulus (cf. the limit to the length of a uniform vertical rod suspended from its upper end being linearly proportional to its strength-to-weight ratio). If supporting its own weight were the only consideration for material selection for the mast of a yacht, compare the advantage of a $[(90/(\pm 45),0)_2]_s$ filament wound glassfiber/epoxy resin tubular mast (axial stiffness 50 GN m^{-2}, density 1.8×10^3 kg m^{-3}) with a traditional spruce mast (Young's modulus 11 GN m^{-2} along the grain, density 700 kg m^{-2}).

First part: To avoid buckling when loaded with a weight W, the length l of a pillar must be less than

$$\tfrac{1}{2}\pi \sqrt{\frac{EAk^2}{W}}$$

where Ak^2 is the moment of inertia of the cross-section of the pillar about an axis through its center at right angles to the plane of buckling. This is Euler's theory of struts. W is the weight of the rod. Assuming that W can be regarded as applied at the mid-point of the rod,

$$l < \pi \sqrt{\frac{EAk^2}{W}}$$

A more accurate analysis,[11] shows that a better equation is

$$l < 2.8 \sqrt{\frac{EAk^2}{W}}$$

Substituting for $W = \varrho g l A$, we find

$$l^3 < 7.84 \frac{Ek^2}{\varrho g}$$

Second part: For the glassfiber mast,

$$\frac{E}{\varrho} = \frac{50 \times 10^9 (\text{N/m}^2)}{1.8 \times 10^3 (\text{kg/m}^3)}$$

$$= 27.8 \times 10^6 \ \text{m}^2\text{s}^{-2} \quad \text{since N} \equiv \frac{\text{m kg}}{\text{s}^2} \ (F = ma)$$

For the spruce mast with identical dimensions,

$$\frac{E}{\varrho} = \frac{11 \times 10^9 \ (\text{N/m}^2)}{7 \times 10^2 \ (\text{kg/m}^3)}$$

$$= 15.7 \times 10^6 \ \text{m}^2\text{s}^{-2}$$

---

[11]J. H. Poynting and J. J. Thomson. *Properties of Matter*, Charles Griffin and Co. Ltd., London, p. 99 (1907).

Hence,

$$\frac{l^3_{glassfiber}}{l^3_{spruce}} = \frac{27.8}{15.7} = 1.77$$

and $l_{glassfiber} = 1.2\ l_{spruce}$ ($\cong 20\%$ advantage).

4. The design specification for the driveshaft for a light truck calls for a fundamental bending frequency (whiplash frequency) of not less than 92 Hz. Show that, to replace the existing 4″ diameter × 6′ long steel driveshaft with a fiber composite tubular driveshaft, an axial specific modulus in excess of $3.3 \times 10^8\,\mathrm{m^2\ s^{-2}}$ is required.

From classical thin shell elasticity theory,[12] the longitudinal frequency of a thin cylindrical shell pinned at its ends and rotating about its axis is

$$\frac{n\left(\dfrac{E}{\varrho}\right)^{1/2}}{2l}$$

$n = 1$ for the fundamental mode.

Investigation of the flexural vibrations of cylindrical shells gives, for the fundamental frequency,

$$f = \frac{\pi}{2} \cdot \frac{R}{l^2} \sqrt{\frac{E}{2\varrho}}$$

R and l respectively are the radius and length of the driveshaft, and E is its axial stiffness.

Re-arranging this equation, we get the specific modulus

$$\frac{E}{\varrho} = \frac{8}{\pi^2} f^2 \left(\frac{l^4}{R^2}\right)$$

$$= 0.8 \times (92)^2 \left(\frac{1}{s^2}\right) \times \frac{(3.35)^4\,(\mathrm{m^4})}{(5.08 \times 10^{-2})^2\,(\mathrm{m^2})}$$

$$= 3.3 \times 10^8\ \mathrm{m^2\,s^{-2}}$$

1.6 Further examples

1. Name three functions served by the plastic in glassfiber reinforced plastic.
2. The ratio Young's modulus to specific gravity, expressed in imperial units, is

---

[12]A. E. H. Love. *A Treatise on the Mathematical Theory of Elasticity 4th edition*, Cambridge University Press, p. 546 (1927).

an unsatisfactory measure of specific modulus on at least three counts. What are they?

3. The modulus of the intermetallic compound CoAl is 300 GN m^{-2} at 250 K falling to 225 GN m^{-2} at 1350 K. CoAl crystallizes with the B2 structure and has a lattice parameter of 0.2860 nm at 293 K. If its coefficient of thermal expansion is 5.6×10^{-6} K^{-1}, what should its specific modulus be at 250 K and 1350 K? In practice, CoAl has a vacancy concentration of $\sim 10^{-1}$ at 1350 K. What is a more likely figure for its specific modulus at this temperature (after A. Wolfenden, Texas A&M University)?

4. The specific modulus of beryllium at low temperatures, and of titanium aluminide (TiAl) at high temperatures, are targets to be reached if fiber reinforced composites are to succeed as high performance materials. The Young's modulus and density for beryllium are 303 GN m^{-2} and 1850 kg m^{-3} respectively. For titanium aluminide they are 173 GN m^{-2} and 3800 kg m^{-3} respectively. Calculate the low and high temperature targets for specific modulus.

5. The following data were obtained on silicon carbide whisker reinforced aluminum alloy 6061 heat-treated to the T-6 condition.

Fiber volume fraction	0	1/5	3/10
Ultimate tensile strength (MN m^{-2})	290	586	793
Young's Modulus (GN m^{-2})	69	121	142
Density (kg m^{-3})	2,700	2,803	2,855

Calculate the specific strength and specific modulus as a function of whisker content.

6. A fiber composite consists of short fibers of glass, aligned in one direction in a matrix of epoxy resin. The proportion of fibers is 50% and they are of length 1 mm and diameter 10 μm. Using order of magnitude values for the elastic moduli involved, calculate the Young's modulus in the fiber axis direction.

7. Assuming a rule of mixtures and perfect adhesion between fibers and matrix, derive an equation for the transverse modulus of a uniaxial fiber reinforced composite. Hence sketch the orientation dependence of Young's modulus for a uniaxial composite.

8. Measurements of engineering strain suggest that carbon fiber becomes stiffer by as much as 20% at 1.5% strain when tested in tension, and more compliant, by as much as 15% at 1.5% axial strain, in uniaxial composites tested in compression. To what extent can such observations be accounted for by converting the engineering strain to true strain?

9. A uniform beam of length l, clamped at one end, bends under its own weight W. Show that the deflection of the free end is given by

$$y(l) = \frac{Wl^3}{8EI}$$

where E is Young's modulus and I is the moment of inertia. Vis-à-vis the space shuttle manipulator arm, compare this deflection with the deflection caused by a force of magnitude W exerted on the free end in a weightless environment (University of Bristol, 1980).

10. Compare the deflections under their own weight of beams of the same material and length in the form of (a) a bar with square cross-section, (b) a circular rod of the same area of cross-section, and (c) a thin-walled tube of the same outside diameter as (b).

11. Derive an equation for the circumferential stress in a rotating circular ring of density ϱ and circumferential speed v. Assuming that this equation can be applied to a filament wound carbon fiber/epoxy resin ultracentrifuge ($\sigma/\varrho = 140$ km) 2 m in diameter, what is the maximum angular velocity?

12. A loaded beam deflects by shear as well as by pure bending. Verify that, for a homogeneous beam, the shear deflection is given by

$$y_d(l) = \frac{3Wl}{AE}$$

A is the area of cross-section of the beam. Hence, show that this deflection is negligible compared with the deflection due to bending for beams long in comparison with their transverse dimensions.

13. Derive the following expression for the difference between free and constrained compliance:

$$\left(\frac{\partial x_1}{\partial F_1}\right)_{F_2} - \left(\frac{\partial x_1}{\partial F_1}\right)_{x_2} = \left(\frac{\partial x_2}{\partial F_1}\right)_{F_2}^2 \left(\frac{\partial F_2}{\partial x_2}\right)_{F_1}$$

where F_1, F_2 are the applied forces conjugate to displacements x_1, x_2.
A certain solid has an isothermal Young's modulus of 1 Mb ($= 10^{11}$ N m^{-2}), a density of 5 M g m^{-3}, a specific heat of 0.8 kJ K^{-1} kg^{-1}, and a coefficient of linear thermal expansion of 2×10^{-6} K^{-1}. What is its adiabatic Young's modulus at 300K (University of Bristol Examination, 1971)?

14. The "unravelled scroll" morphology of graphite whiskers is attributable to the so-called Eshelby twist imparted to a whisker by a screw dislocation lying along its axis. Summarise Eshelby's reasoning for the surprising fact that a single axial screw dislocation is a stable configuration, and read and comment upon F. C. Frank's theory that multiple axial screw dislocations are also stable configurations and can, therefore, account for axial displacements of magnitude very much larger than one Burgers vector. References: J. D. Eshelby, "Screw Dislocations in Thin Rods," J. Appl. Phys., 24:176–179 (1953). F. C. Frank, "Spontaneous Breaking of Chiral Symmetry in the Growth of Crystal Whiskers: An Origin for Giant Screw Dislocations," Phil. Mag., 56:263–268 (1987).

2

Materials and processing

The variety of fibers used in modern composites includes: cellulose, three forms of asbestos, five glassfiber compositions, tungsten core and carbon core variations of boron fiber, at least three microstructurally different kinds of carbon fiber, two aramid fibers available from three different manufacturers, two commercially available alumina fibers and two kinds of silicon carbide fiber. The choice of matrix materials is no less daunting, with some fourteen thermosets, five thermoplastics, the metals, glasses and ceramics all widely available.

2.1 Structural, crystalline versus non-crystalline, classification of materials

The classification of polymers as either thermoplastic or thermosetting is a manufacturer's distinction. A much more important division is into the crystallizing and the non-crystallizing polymers. Polystyrene is thermoplastic, and bakelite is thermosetting; both are non-crystallizing. Polyethylene and nylon are crystallizing; the percentage crystallinity of a polyethylene bag, for example, is 50–60%.

All metals are crystalline. The individual crystals can sometimes be seen with the naked eye, for example on bronze door knobs which, over the years, have been etched by the acid from human hands. Cast metals are another example where the boundaries between crystals (grains) can sometimes be seen by the unaided eye.

Wood is about 50% cellulose, which is crystalline.

Glasses are *not* supercooled liquids. Materials vary in the extent to which they can be supercooled. Figure 2.1 shows the changes that occur in any thermodynamic property (heavy line) during cooling of say, glycerol or toluene at, say, 10^{-2} K s^{-1}.

A somewhat similar curve is obtained for any material which does not spontaneously produce nuclei of solid phase. The pure elements, pure aluminum for example, can likewise be supercooled, by as much as 20% of the absolute melting temperature. In the case of aluminum, titanium carbide present as trace impurity inclusion reduces the supercooling by 2 to 3 K. Suppose a pinch of TiC crystals were

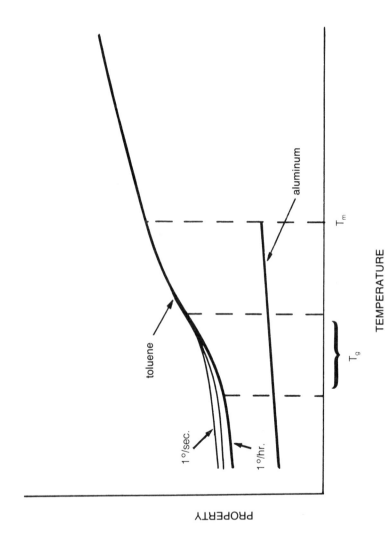

Figure 2.1. Property changes during supercooling of liquids. The property may be any thermodynamic property, for example $1/\varrho$ where ϱ is the density, or S the entropy, or $1/n$, where n is the refractive index, or ϵ the dielectric constant.

dropped into pure liquid aluminum. If we introduce the crystals just below the melting temperature T_m freezing immediately takes place, that is the aluminum completely solidifies and the slope of our property versus temperature relationship (the slope is a coefficient of thermal expansion if the property measured is $1/\varrho$, or a specific heat divided by absolute temperature if the property is S, etc.) abruptly changes to that characteristic of solid aluminum. This is a first-order transition and is characterized by a latent heat. In the absence of any nucleating crystals, what we have below T_m is supercooled liquid for which the slope of the property versus temperature curve changes smoothly as sketched in Figure 2.1. What happens on cooling further may depend on the rate of cooling. With inorganic glass-forming liquids we have a "fan" of property-temperature relationships, as shown in Figure 1.1, one relationship for each of the cooling rates 1 K per second, 1 K per minute, 1 K per hour. The molecular rearrangement responsible for the viscosity change during glass transition is not the same as that which takes place during solidification to crystalline material. What single temperature to take as the glass transition temperature (T_g) is disputable. One practical measure for T_g is the temperature defined by the intersection of projections of the adjacent linear segments of the property-temperature relationship recorded at a standard cooling rate of 1 K per minute. It is evident that the material becomes a glass in a temperature range where properties depend on the thermal history. Note that the material is a supercooled liquid only above T_g.

Glass is brittle because the re-adjustment of relative positions of molecules responsible for the relief of stress at stress concentrations is also responsible for flow when in the liquid state. That is, flow and relaxation of shear stresses are by the same molecular mechanisms. Hence, at room temperature, sharp cracks in glass remain sharp and can grow. So, brittleness, and high strength if no cracks are present, are characteristic properties of glass.

Crystalline materials can yield in a plastic manner. What happens when the limit of elasticity is exceeded is sketched as a curve in the stress (σ) versus strain (ϵ) relationship (solid line) shown in Figure 2.2. However, plotting a stress-strain curve assumes that time effects are unimportant. In particular we are ignoring creep.

To solve real problems in plasticity, the stress-strain relationship is usually idealized by assuming no work hardening (denoted in Figure 2.2 by the horizontal line after the yield point). That is, we ignore work hardening. Such an idealized material is called a plastic-elastic material.

Now any yielded metal having stress in it, contains elastic and plastic strains. We shall ignore the former. That is, we shall make a further simplification and assume a plastic-rigid material (infinite elastic modulus)—broken line stress versus strain relationship in Figure 2.2.

How valid are the above assumptions for the prime candidate metals for metal matrix composites, aluminum and magnesium? The assumption of no strain hardening is acceptable for aluminum, but is not a good assumption for magnesium. The assumption of a plastic-rigid material is also not a good one. It is appropriate in metal fabrication processes like rolling, extrusion and wire-drawing, but in a rigid as-cast fiber reinforced metal, where we effectively have an internal

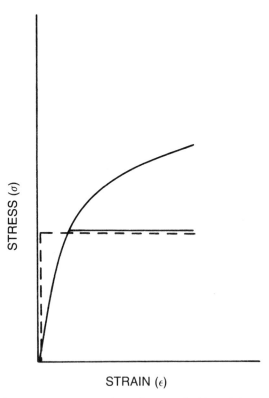

Figure 2.2. Showing stress (σ) versus strain (ϵ) for a metal with work hardening (solid curve), without work hardening (horizontal line), and for a plastic-rigid material (broken lines).

pressure in a thick-walled metal tube generated by thermal expansion mismatch [refer to Figure 2.3(a)], the plastic strains cannot exceed the order of magnitude of elastic strains. In fact, whenever we have a boundary between elastically and plastically deformed material, we have a zone of about the same elastic and plastic strains, see Figure 2.3(b); therefore, neglecting the elastic strains cannot be justified.

In Figure 2.2, only one component of stress is operating. To generalize to triaxial states of stress, consider in Figure 2.4(a) three coordinate stress axes, σ_1, σ_2, σ_3. The yield stress in uniaxial tension along the coordinate axes is denoted by a filled circle on each stress axis. What if we choose a radius from the origin other than along a coordinate axis? This leads us to the concept of a yield surface enclosing the origin. What is the shape of this surface? Well, we resort to the experimental fact that, to a first approximation, yielding of a polycrystalline metal is unaffected by moderate hydrostatic pressure (say of magnitude up to 100 times the shear strength) either acting alone or superposed on other stresses. Hydrostatic pressure is represented by a line making equal angles with the coordinate axes, that is by the [111] direction, coming out of the plane of Figure 2.4. The yield surface is, then, open in this direction. The same is true, that the yield surface is open along [111], if we add hydrostatic pressure to a uniaxial yield stress. The yield surface must, therefore, be a cylinder with generators parallel to [111]. It must also be symmetric in σ_1, σ_2, σ_3 since these are accidents of labelling.

Von Mises (and Huber and Maxwell) suggested that we should take as yield surface a circular cylinder. Viewed down the [111] direction looking toward the origin, this is seen as a circular section, Figure 2.4(b). Tresca (and Mohr) point out that yielding occurs when the maximum shear stress reaches a critical value. The maximum shear stress equals one half the difference between the two extreme principal

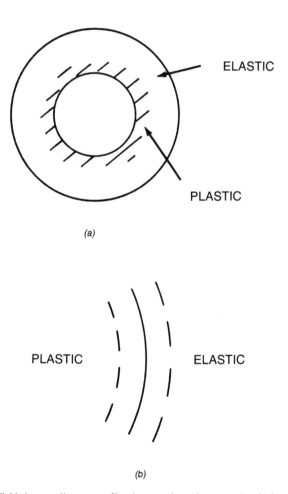

Figure 2.3. (a) Yielded zone adjacent to a fiber in a metal matrix composite, (b) boundary between elastically and plastically deformed metal.

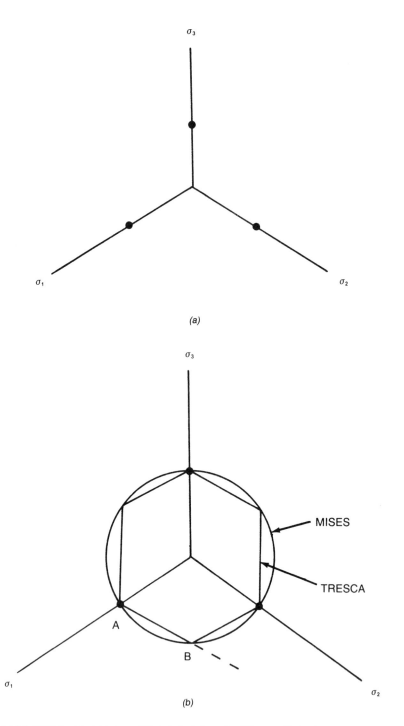

Figure 2.4. (a) Uniaxial yield stress (filled circles) for each of the coordinate stress axes σ_1, σ_2, σ_3, (b) Mises and Tresca yield surfaces for triaxial states of stress.

stresses so, taking $\sigma_1 >> \sigma_2 >> \sigma_3$, our condition for yield is that

$$\sigma_1 - \sigma_3 = \text{a constant} \tag{2.1}$$

On the $\sigma_1\sigma_2$ plane, $\sigma_3 = 0$, which is as small as it can be. Moving along line AB in Figure 2.4(b), σ_1 is constant, so $\sigma_1 - \sigma_3$ is constant. Beyond B, σ_2 is the largest principal stress and the condition for yield is $\sigma_2 - \sigma_3 = $ a constant. Hence the Tresca yield surface is a cylinder whose section is a hexagon.

Experimentally, it is found that the Mises is the better of the two criteria. Mathematically, the Mises criterion is

$$(\sigma_1 - \sigma_2)^2 + (\sigma_2 - \sigma_3)^2 + (\sigma_3 - \sigma_1)^2 = 6k^2 \tag{2.2}$$

where k is the maximum shear stress measured in pure shear tests.

One other class of material used in some fiber composites, laminated tires for example, is the gels or rubbers. Natural rubber is one of the crystallizing polymers; if cooled to $-10°C$ or so, or even at ordinary temperatures under stress, natural rubber will crystallize.

When behaving like a rubber, that is, when exhibiting rubber elasticity, these materials behave differently from all other materials in the following important respect. When more conventional materials, steel for example, are stretched, they get cooler. However, when rubber is stretched it gets hotter. Whereas in steel, the applied stress does work against forces of interatomic attraction, in rubber it does work against entropy, depressing the entropy of the material. We freeze the random motions of the molecules when a rubber solidifies and have a glass transition temperature.

So, on the one hand, the gels and rubbers form a set with all other crystalline materials, and, on the other hand, they form a set with the inorganic glasses, see Figure 2.5. For the sake of completeness, the liquid crystals, the mechanical behavior of which is that of the wet soap that accumulates in the base of a soap dish, are also included in Figure 2.5.

Returning to Figure 2.1, the property-temperature changes recorded on cooling are reversible, except that variations occur near the transition range. Thus, if we quench and reheat a glass, a slight sag in the curve is observed near the change in slope, and the sag can be negative. This means that we can have negative specific heat, that is, heat is given out as we slowly raise the temperature of our glass.

The glass transition temperatures for plastics are *not* second-order transition temperatures. T_g is fairly well defined but gives rise to a fan of multi-valued property of the kind shown in Figure 2.1. For some materials, a second transition to a fan of multi-valued property occurs at lower temperatures. Plastic sulphur, a non-carbon polymer, is one material with two such T_g's. Most of the electrostatic polarity is in the side groups of long chain molecules, as shown in Figure 2.6(a), and it is re-orientation of the side groups that reveals the first T_g in the dielectric constant (ϵ) measured at very low frequency, see Figure 2.6(b). The freedom of movement of the groups is multi-valued. The second T_g, Figure 2.6(b), is asso-

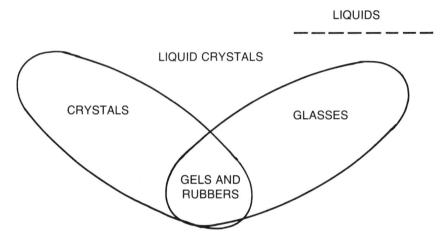

Figure 2.5. Sets for materials classified on a structural basis.

ciated with re-orientation in the side groups and re-orientation of the backbone structure. The final decrease in ϵ is due to near randomness at high temperature. As a consequence of this, we need a bigger field in order to get a given orientation.

2.2 Technological polymers

Thermoplastics are chain polymers. Bifunctional units, that is, monomers each with a group capable of reacting at its ends, link up to form a chain molecule. Chain polymers are soluble and fusible, and are therefore intrinsically crystallizable.

Thermosets are network polymers. Multi-functional monomers link together to form a three-dimensional network. Network polymers are insoluble, infusible, and non-crystallizable.

Curing (of resins) and *vulcanizing* (of rubber) are examples of network formation derived from latent functionality. Trifunctional units for example, for which two of the functional groups are more reactive than the third, could form a long chain along which we have unreacted functional groups. Subsequent reaction of these groups later creates a network, and so causes the polymer to be permanently set.

Polycondensation. There are two kinds of polymerization reactions, polycondensation and additive polymerization. In the case of polycondensation, two monomers come together and a molecular unit, usually water, splits off allowing formation of the link between the units.

Additive polymerization is polymerization without any splitting off of molecular units. The polymerization reaction has to be activated, and this is done by introducing free radicals or ions or both.

Polyesters form by a condensation reaction:

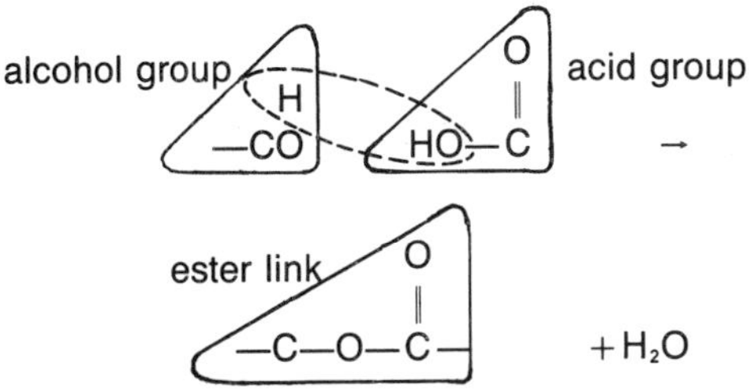

The unsaturated bond in the ester link is capable of forming a cross-link with a similar polyester molecule either directly, or (more usually) via a monomer such as styrene.

Epoxies – Diglycidyl ether of bisphenol A (DGEBA), for example, can be manufactured by reacting epichlorohydrin with bisphenol A.

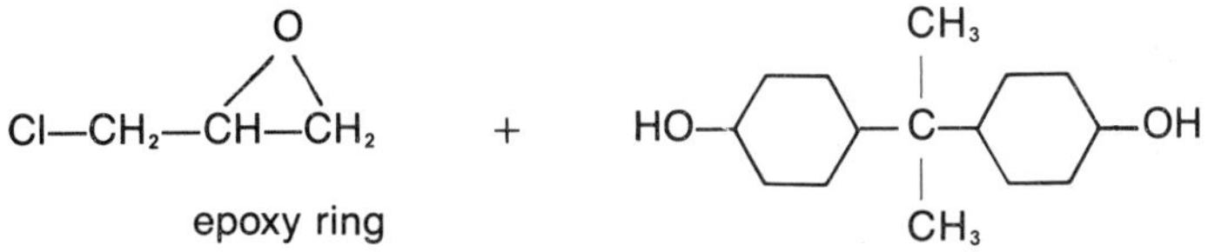

In the presence of excess epichlorohydrin and caustic soda, a chlorhydrin intermediate is formed:

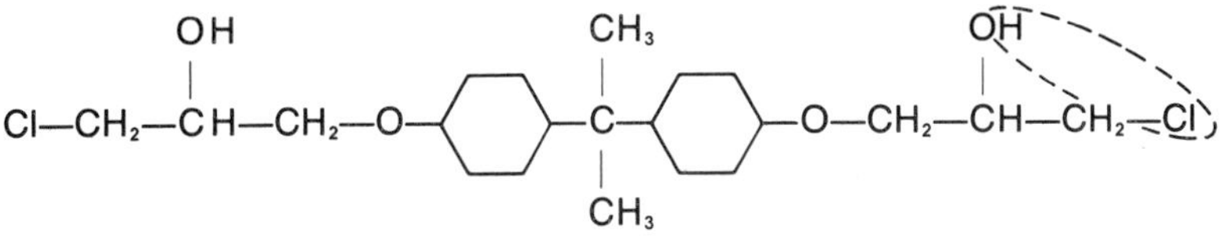

The caustic then acts as a dehydrohalogenating agent to re-form the epoxide rings and neutralize the hydrochloric acid:

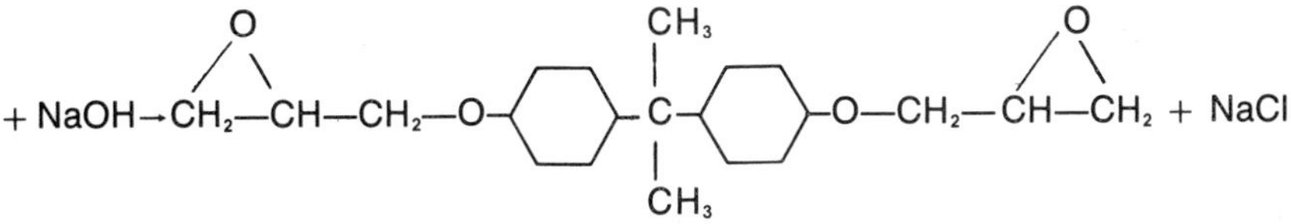

Cross-links with similar molecules can then be formed by additive reactions with amines:

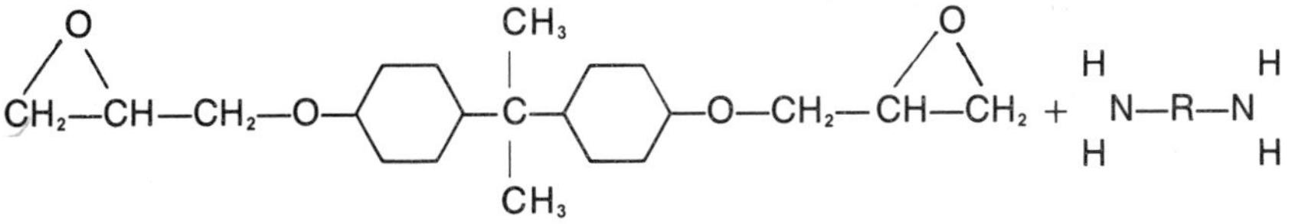

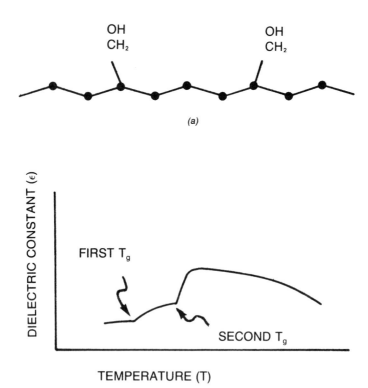

(a)

(b)

Figure 2.6. (a) Illustrating side groups on a polymer chain, (b) low frequency dielectric constant (ϵ) versus temperature (T) for a polymer that undergoes two glass transitions.

leaving three functional groups for reaction with an epoxide ring belonging to a neighboring molecule.

2.3 Manufacture of prepregs

Pre-impregnated material (prepreg) consists of a parallel array of rovings or tows, usually spread out into a sheet, and pre-impregnated with liquid matrix

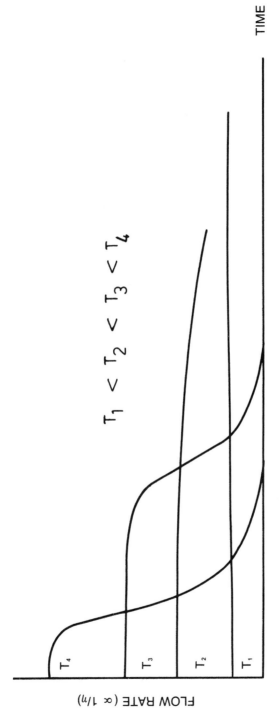

Figure 2.7. Isothermal flow rate curves for resin drainage during prepreg manufacture.

material (polymer or metal). In the case of thermoset matrix material, the resin is partially cured (B-stage) to a tacky condition so that the prepreg can be handled when laying up laminates. In the case of a metal matrix prepreg, the metal is solidified. The thickness of the unidirectional tape so obtained is typically 120 μm for epoxy matrix prepregs.

Sheet molding compounds (SMC) are short-fiber (random) reinforced resins that have been "filled" with inert particles such as chalk so as to impart rheological characteristics that ensure flow under pressure.

There are two countervailing effects which determine the rate at which resin flows during prepreg manufacture. The immediate response to increases of temperature is lowering of the viscosity, which is short-lived, since curing, i.e., cross-linking, gives rise to an increase of the resin viscosity. The effects on flow rate are sketched schematically in Figure 2.7.

The area under each curve in Figure 2.7, i.e., the integral of the flow rate versus time relationship at a given temperature, is a measure of the amount of resin which did flow at that temperature. This is a maximum at some temperature between room temperature and the final cure temperature and is the optimum temperature for prepreg preparation.

The fabrication of laminates from modern prepregs is reckoned not to involve flow and drainage of resin on a scale large enough to alter the fiber volume fraction from that delivered by the prepreg manufacturer. The fiber volume fraction is typically 3/5 [cf. $\pi/(2\sqrt{3})$ for fiber close packing].

Poly(p-phenylene sulfide), a thermoplastic, can be spun into 25 μm diameter yarn and comingled with carbon fiber to produce prepreg yarn, which can then be woven into fabric. The thermoplastics in general are also easily powdered and packed between fibers to make powder preforms.

2.4 Glossary of fabrication methods

Hand lay-up into a mold previously coated with release agent is widely used for "one offs."

Spray-up process in which liquid resin and chopped fibers are simultaneously sprayed from a gun lends itself to continuous production of low technology composites.

Press molding, between matching male/female dies in the case of sheet molding compounds and between parallel platens in the case of laminates, is normally performed hot (resin viscosity $\sim$ 10^2 N m^{-2} s) and yields close tolerances. Some thermoplastic matrix laminates for high quality electrical and mechanical components such as radomes are manufactured by a press molding process known as film stacking, in which lightly impregnated fabric reinforcement prepreg is interleaved with thermoplastic layers, and the stack is bonded together at high temperature and pressure.

Vacuum bag molding (Figure 2.8) exploits the pressure of the atmosphere (10^5 N m^{-2}) to secure contact with the mold, to consolidate the lay-up in the case of laminates, and to expel air and volatiles during curing.

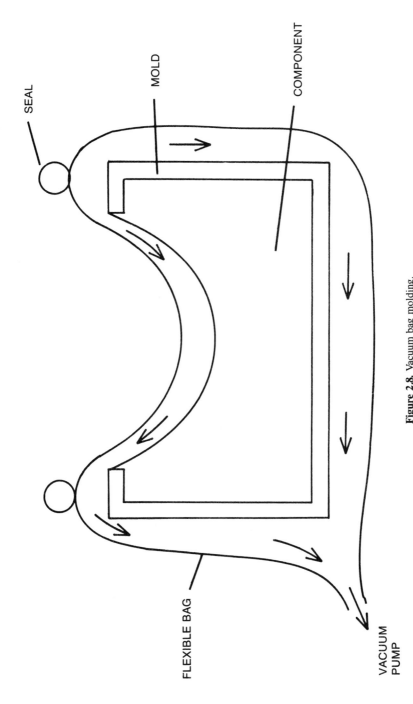

Figure 2.8. Vacuum bag molding.

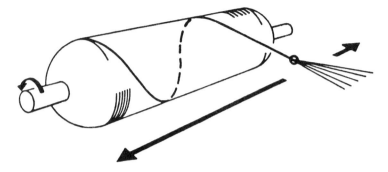

Figure 2.9. Helical (lathe-type) filament winder.

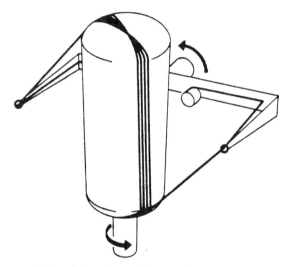

Figure 2.10. Polar (whirling arm) filament winder.

Figure 2.11. Illustrating relative fiber displacements during uniaxial compression molding.

Autoclave molding provides for consolidating pressures greater than one atmosphere. The component, laid up on a mold, is enclosed in a flexible bag inside an autoclave (pressure vessel). After evacuating the bag to remove air and volatiles, an overpressure of typically 1.5 MN m^{-2} is applied to the bag by steam or other gas. There is usually provision for raising the temperature through the resin cure cycle.

Resin transfer molding (also known as liquid molding) starts with the fibers assembled as a preform, and enclosed in a mold into which the resin is injected, usually at a viscosity of $\cong$ 10^4 N m^{-2} s. Radomes, for example, are conveniently made by injection of liquid resin into an array of pre-positioned woven glassfiber stockings.

Filament winding is the winding of a continuous band of resin-impregnated scorings (fibers) onto a mandrel. The mandrel is usually of constant circular cross-section. Two main types of winder, helical and polar, are sketched in Figures 2.9 and 2.10.

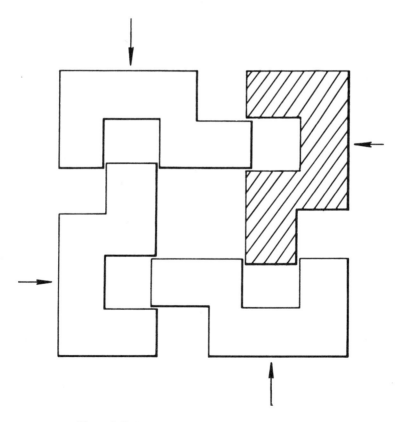

Figure 2.12. Bi-dimensional compression molding device.

For filament winding to be stable, that is, to avoid any possibility of fiber slippage, it is necessary to identify geodesic paths. A geodesic path is the shortest distance between two points on a surface. Polar winders wind over end and are thus able to form closed-end cylindrical tubes. Filament winding lends itself to the fabrication of very high fiber volume fraction $\eta = 4/5$ composites, cf. $\eta = \pi/(2\sqrt{3})$ for fiber close packing.

Pultrusion is a non-stop process for manufacturing uniaxial continuous fiber reinforced components. The fibers are used to pull the composite through a die; continuous rovings, fed from a creel, are pulled into a resin bath and then through a heated die inside which curing takes place. In Chapter 1, it was pointed out that, to realize the full potential of composite materials, it will be necessary to exploit highly anisotropic lay-ups. Pultrusion is a very efficient process for manufacturing highly anisotropic, constant cross-section components. As a method for mass production of fiber reinforced plastic rods and tubing, it has not yet been fully exploited – due to the difficulty of joining to other components. One inexpensive solution to the problem is "crimp-bonding" of end-fittings. Here a loose-fitting sleeve is clinched around the pultruded length of composite, leaving a very small gap which is automatically filled with adhesive.

Pullforming, or interrupted pultrusion, is a refinement widely used for mass production of short-length, uniaxially reinforced composites.

Bi-dimensional compression molding comes closer than any other process to ultra-high fiber content composite manufacturing. The problem with a unidirectional squeeze is that it does little to improve fiber packing. As depicted in Figure 2.11, when squeezed vertically, the two lower fibers move apart to allow approach by the upper fiber. To consolidate the array, it would be necessary to simultaneously squeeze horizontally and vertically. Figure 2.12 shows a plurality of interfitting G-shaped anvils capable of imparting bi-dimensional compression. Fiber loadings very close to the ideal packing can be realized with bi-dimensional compressional molding.

2.5 Three-dimensional weaves

There exist an infinite number of weaves with cubic symmetry. Figure 2.13 shows an orthogonal weave of straight rods arranged along the principal directions x, y, z. By making the rods of square cross-section, it is seen that they pass three at a point and, in each unit cell, they define two cubic cavities with edge length equal to the fiber thickness. Thus the fiber volume fraction is $\eta = 3/4$. Bevelling of the fiber edges, so that they approximate to circular cross-section, only slightly lowers the fiber volume fraction.

Orthogonal weaves are not reinforced against shear. To a first-order approximation, no member in an array of wires arranged as shown in Figure 2.13, for example, is stretched by shear in a cube plane. For this reason, other three-dimensional weaves are more important in mechanical applications.

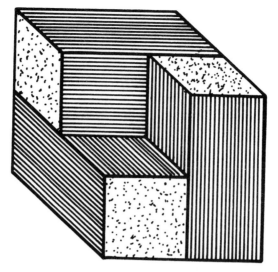

Figure 2.13. < 100 > orthogonal weave.

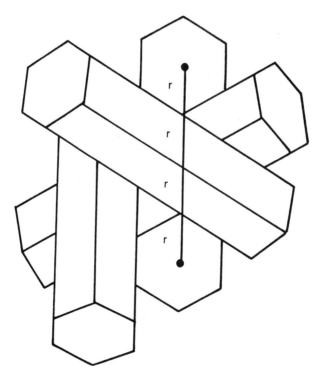

Figure 2.14. Projection along one of the fiber directions of a < 111 > three-dimensional weave.

Figure 2.14 shows a $<111>$ weave of hexagonal prisms. With {112} prism faces, they pass in pairs and they define one tetrahedral cavity per unit volume, with edges parallel to the six $<110>$ directions and with edge length equal to $r\sqrt{6}$, where r is the hexagon edge length. The distance between parallel prisms is $4r$. A sphere of radius $2r$ centered on the cavity includes one fiber in each of the four orientations. The volume fraction of this sphere taken up by the tetragonal cavity is

$$\frac{\frac{\sqrt{2}}{12}(r\sqrt{6})^3}{\frac{4}{3}\pi r^3} \sim \frac{\sqrt{3}}{4}$$

Hence the fiber volume fraction for this weave is $\eta = 1 - (\sqrt{3}/4)$.

Another cubic symmetry weave worth considering is the $<110>$ weave, for which there are six fiber orientations. Also worth considering are combinations, involving $<110>$ and $<111>$ for example.

Most practical three-dimensional fiber reinforcements are characterized by fiber volume fractions significantly smaller than the ideal values for rigid cords. Smaller fiber volume fractions permit bending of fibers to the high curvatures demanded by the weft and warp in traditional textile technology. In two-dimensional cloth, the warp are the lengthwise threads and the weft (or woof) are the threads at right angles to the warp. New terminology is required in order to describe three-dimensional weaves. One version, borrowed from crystallography, is to regard each fiber direction as a symmetry axis and to define it by the symbol representing that symmetry, together with a fractional number to keep count of the occupancy of axes belonging to each family of axes. Thus, the weave shown in Figure 2.13 is □ 3/3, and that shown in Figure 2.14 is △ 4/4.

Once the three-dimensional array of fibers has been put in place, it is impregnated with matrix material. In the case of three-dimensionally reinforced carbon/carbon composites, this is usually accomplished by either liquid infiltration or vapor deposition of organic material which is subsequently carbonized by heat treatment. To minimize entrapment of air- or vapor-filled voids, the matrix is gradually built up by carrying out a number of infiltration/carbonization operations. Carbon/carbon composites have exceptional wear resistance and erosion resistance (see Chapter 8, Section 8.8).

Another important class of three-dimensionally woven fiber composites is the self-lubricating, anti-friction/anti-wear materials used as liner material between the inner and outer races of bearings. Polytetrafluoroethylene (PTFE) is the classic dry bearing material. However, PTFE has low modulus and strength, it is easily deformed, and it cannot be easily bonded to other materials. By using PTFE in the form of fibers interwoven in a three-dimensional weave with conventional fibers, these limitations can be overcome. Figure 2.15 illustrates one such weave.

Figure 2.15. Self-lubricating, three-dimensionally reinforced composite. Threads which outcrop the upper surface are PTFE (shaded). All others are reinforcing fibers.

2.6 Sintering of ceramic matrix composites

Since plastic deformation by shear processes (dislocation glide, grain boundary sliding, mechanical twinning) cannot occur under hydrostatic pressure, the particle shape change, from quasi-spherical to space-filling polyhedral, that takes place during isostatic pressing is brought about by the movement of atoms or, more usually, ions. There are several processes at work, including stress-induced dissolution and stress-induced melting. High elastic strain energy density material, at compressive contacts between particles of powder, preferentially dissolves in the fluid noncrystalline material known to exist between grains in polycrystalline ceramics, and is redeposited on less highly stressed solid material. The second process, thermodynamic melting, occurs for a solid that contracts on melting. Silicon and ice are classic examples. The dependence of melting temperature on pressure is given by Clapeyron's equation:

$$\frac{dT}{dp} = \frac{\Delta V}{\Delta S}$$

(2.3)

where ΔV is the volume change, p the confining pressure and ΔS the associated entropy change.

Sintering is also achieved by diffusion down concentration gradients. This may be due, in part at least, to variations in vacancy concentration attributable to stress, given that vacancy concentrations are higher in tensioned than in compressed material. However, diffusion in ceramics is slow, mainly due to the coupling between migration of different species that is necessary if stoichiometry is to be maintained. Typical diffusion coefficients (D) are $10^{-16} m^2 s^{-1}$ (10^{-12} cm^2 s^{-1}) for Al^{3+} and $10^{-18} m^2 s^{-1}$ (10^{-14} cm^2 s^{-1}) for O^{2-} in alumina at 1700 °C; using, as order of magnitude estimate for the time (t) for diffusion through distance (x), the approximate Fickian diffusion formula

$$t = \frac{x^2}{D}$$

we find

$$t = \frac{10^{-4}(\text{m}^2)}{10^{-18}(\text{m}^2/\text{s})} = 10^{16} \text{ seconds}$$

$$\sim 3.2 \times 10^8 \text{ years}$$

for oxygen diffusion through 100 mm of charge at 1700°C.

Atomic transport in ceramics also occurs in electric fields, and in electromagnetic fields. The former makes possible so-called plasma sintering, and the latter makes possible microwave sintering.

When fabricating fiber reinforced ceramics, the physical and chemical properties of the fiber often dictate which fabrication process best suits the overall composite. Not least among the fiber properties to be considered when choosing the fabrication process is thermal conductivity. The heating time (t) for a cross-section of fiber is of the order

$$t = \frac{r^2}{\alpha} \qquad (2.4)$$

where r is the fiber radius and α its thermal diffusivity.

$$\alpha = \frac{\varkappa}{\varrho\, c_p} \qquad (2.5)$$

where $\varkappa$ is the thermal conductivity, ϱ the density and c_p the specific heat capacity of the fiber material. Taking as trial fiber material, titanium diboride (TiB_2), $\varkappa = 65\,\text{W m}^{-1}\,\text{K}^{-1}$ at red heat ($\cong 600°C$), $\varrho \cong 2 \times 10^3 \text{kg m}^{-3}$, $c_p \cong 10^3\,\text{J K}^{-1}\,\text{kg}^{-1}$. Hence

$$\alpha = \frac{65 \ (\text{J/m s K})}{2 \times 10^3 \ (\text{kg/m}^3) \times 10^3(\text{J/K kg})}$$

$$= 3.25 \times 10^{-5}(\text{m}^2/\text{s})$$

so, for $r \cong 18\ \mu\text{m}$,

$$t = \frac{3.24 \times 10^{-10}(\text{m}^2)}{3.25 \times 10^{-5}(\text{m}^2/\text{s})}$$

$$\cong 10^{-5}\text{s}$$

On the one hand a 10 μs heating time ensures that sintering of TiB_2 short fiber reinforced ceramics introduces no significant temperature gradient, and hence no thermal shock, in the fibers; on the other hand, the fact that all of the fiber is uniformly raised to the sintering temperature means that thermal decomposition is not confined to the fiber surface. One notable system where degradation of the fiber during sintering is a problem is silicon carbide fiber reinforced alumina; prolonged sintering causes the SiC to oxidize and hence combine with the Al_2O_3 to form mullite [$Al(AlSiO_5)$].

2.7 Worked examples

1. Infiltration of fibers by liquid resins and by liquid metals is basic to many fiber composite manufacturing processes. Consider the flow of liquid of viscosity η through a cylindrical channel of length l and radius a, along the axis of which is a fiber of radius b. If the flow is laminar and is driven by a pressure difference p between the ends of the channel, find the volume of liquid flowing in unit time (University of Bristol, 1982).

The flow, sketched in Figure 2.16, is described by the Navier-Stokes law:

$$\varrho \, \frac{D\underset{\sim}{u}}{Dt} = \varrho \underset{\sim}{f} - \nabla P + \eta \nabla^2 \underset{\sim}{u} \tag{2.6}$$

where ϱ is the density of the liquid, and $\underset{\sim}{u}$ its velocity, and $\underset{\sim}{f}$ is a body force such as gravity.

$$\underset{\sim}{u} = (0, 0, u(r)) \tag{2.7}$$

or, referred to Cartesian coordinates,

$$\underset{\sim}{u} = (u, v, w) \tag{2.8}$$

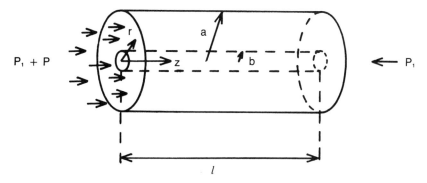

Figure 2.16.

$$\left(\frac{D\underset{\sim}{u}}{Dt}\right)_x = \frac{\partial u}{\partial t} + u\frac{\partial u}{\partial x} + v\frac{\partial u}{\partial y} + w\frac{\partial u}{\partial z} \qquad (2.9)$$

More generally, for vector $\underset{\sim}{F}$,

$$\frac{D\underset{\sim}{F}}{Dt} = \frac{\partial\underset{\sim}{F}}{\partial t} + u \cdot \nabla\underset{\sim}{F}$$

If we assume that the liquid is incompressible,

$$\nabla\underset{\sim}{u} = 0 \qquad (2.10)$$

if we further assume steady flow,

$$\frac{\partial\underset{\sim}{u}}{\partial t} = 0 \qquad (2.11)$$

Hence, Equation (2.9) becomes

$$\frac{D\underset{\sim}{u}}{Dt} = 0 \qquad (2.12)$$

Returning to Equation (2.6), for the flow considered here,

$$\eta\nabla^2\underset{\sim}{u} - \nabla P = 0$$

That is,

$$(\nabla^2 u_r, \nabla^2 u_\theta, \nabla^2 u_z) = \left(\frac{\partial P}{\partial r}, \frac{1}{r}\frac{\partial P}{\partial\theta}, \frac{\partial P}{\partial z}\right)$$

Now

$$\nabla^2 u_r = \nabla^2 u_\theta = 0$$

so,

$$\frac{\partial P}{\partial r} = \frac{\partial P}{\partial\theta} = 0$$

and

$$P = P(z)$$

We are left with only the z part,

$$\nabla^2 u_z = \frac{\partial P}{\partial z} \qquad (2.13)$$

comparing this with the wave equation

$$\nabla^2 \psi = \frac{1}{r} \frac{\partial}{\partial r} \left(r \frac{\partial \psi}{\partial r} \right) + \frac{1}{r^2} \frac{\partial^2 \psi}{\partial \theta^2} + \frac{\partial^2 \psi}{\partial z^2}$$

we get

$$\frac{1}{r} \frac{\partial}{\partial r} \left(r \frac{\partial u(r)}{\partial r} \right) + 0 + 0 = \frac{\partial P}{\partial z}$$

$$= A, \text{ a constant}$$

Hence

$$P = A_z + C \qquad (2.14)$$

and

$$\frac{\eta}{r} \frac{\partial}{\partial r} \left(r \frac{\partial u}{\partial r} \right) = A$$

Therefore

$$\underline{u}(r) = \frac{r^2 A}{4\eta} + B\ln r + D \qquad (2.15)$$

The boundary conditions

$$P(o) = P_1 + P; \quad P(l) = P_1; \quad u(a) = u(b) = 0$$

give A, B, C, D. Hence

$$u(r) = \frac{P}{4\eta l} \left\{ a^2 - r^2 - \frac{(a^2 - b^2)\ln\dfrac{a}{r}}{\ln\dfrac{a}{b}} \right\} \qquad (2.16)$$

The volume flow is

$$Q = \iint \underline{u} \cdot d\underline{S}$$

$$= \int_b^a dr \int_0^{2\pi} d\theta \cdot r u(r)$$

$$= \frac{\pi P}{8\eta l} \left\{ a^4 - b^4 - \frac{(a^2 - b^2)^2}{\ln \dfrac{a}{b}} \right\}$$

(2.17)

Note that, for a fiber, $b \to 0$ and the volume flow

$$Q \to \frac{\pi P a^4}{8\eta l}$$

(2.18)

even though the velocity is zero along the center, see Figure 2.17. Equation (2.18) describes the law discovered by Poiseuille for the flow of liquids through capillary tubes.

2. A mandrel of diameter d_1 is wound with fiber of diameter d at tension T until the outer diameter is d_2. Derive an approximate formula for the total pressure on the mandrel (University of Bristol, general paper, 1986).

Use this expression to estimate the bursting pressure of a 1 m inside diameter, 10 mm wall thickness 90° aramid fiber filament wound pressure vessel. The fiber diameter is 12 μm and its tensile strength is 2.3 GN m^{-2}.

Let the pressure be P. In Figure 2.18, consider length dl of one turn of glass-fiber.

$$dP \, d \, dl = 2T \sin \theta$$

$$\sin \theta = \frac{dl/2}{r}$$

therefore

$$dP = \frac{T}{r} \frac{1}{d}$$

Now a layer at radius r increases the radius by $dr = d$, so

$$\frac{dP}{dr} = \frac{T}{rd^2}$$

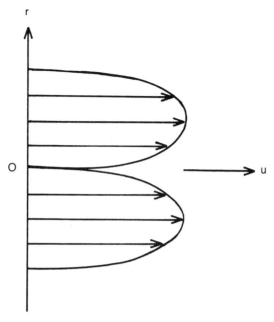

Figure 2.17. Distribution of velocity of liquid flowing through a channel that contains a stationary fiber along its axis.

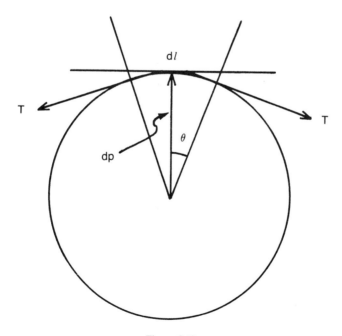

Figure 2.18.

hence

$$P = \int_{r_1}^{r_2} \frac{T}{d^2} \ \frac{dr}{r}$$

$$= \frac{T}{d^2} \ln \left(\frac{r_2}{r_1} \right)$$

$$= \frac{T}{d^2} \ln \left(\frac{d_2}{d_1} \right) \qquad (2.19)$$

Second part:

$$\frac{T}{d^2} \ln \left(\frac{d_2}{d_1} \right) = \frac{T}{d^2} \ln \left\{ 1 + \frac{d_2 - d_1}{d_1} \right\}$$

$$= \frac{T}{d^2} \left\{ \frac{d_2 - d_1}{d_1} - \frac{(d_2 - d_1)^2}{2d_2} + \cdots \right\}$$

$$\cong \frac{T}{d^2} \left(\frac{d_2 - d_1}{d_1} \right)$$

The fiber tensile strength

$$\sigma = \frac{T}{\pi \left(\dfrac{d^2}{4} \right)}$$

Hence the bursting pressure is

$$P = \frac{\pi \sigma}{4} \ \frac{(d_2 - d_1)}{d_1}$$

$$= \frac{\pi \times 2.3 \times 10^9 (\text{N/m}^2)}{4} \times \frac{0.005(\text{m})}{1 \ (\text{m})}$$

$$= 9 \, \text{MN m}^{-2}$$

2.8 Flowing matrix

Poiseuille flow of liquid matrix material past a stationary fiber (Section 2.7, question 1) raises the question of how much stress is imparted to the fiber. In the case of a highly viscous flowing matrix such as creeping metal, there is the possibility of fiber fracture. Following A. Kelly and W. R. Tyson,[13] the axial stress $d\sigma$ in an incremental length dx of fiber due to a shear stress τ, exerted by a plastically flowing matrix is

$$d\sigma = \frac{2\pi r\,\tau dx}{\pi r^2} \qquad (2.21)$$

where r is the fiber radius.

Integrating Equation (2.21) over the fiber length l, we get

$$\sigma = \frac{2\tau}{r} \int_o^l dx$$

$$= 2\tau\, \frac{l}{r} \qquad (2.22)$$

In the context of a creeping short fiber reinforced metal, the flow field sketched in Figure 2.16 would be imaged in the plane perpendicular to the fiber axis passing through the origin O at the fiber center.

Hence l in the above analysis represents only one half of the actual fiber length. If σ is the breaking strength of the fiber, it is evident that there exists a critical fiber length $l_{critical} = (\sigma/\tau)r$ to which longer length fibers will fracture during creep of the metal matrix. $l_{critical}/r = \sigma/\tau$ is known as the critical aspect ratio for fibers used in the fabrication of metal matrix composites and is of the order of 100:1.

2.9 Directionally solidified eutectics

Most eutectic phase mixtures grow with either sheet-like or rod-like morphology. The latter closely resembles the microstructure required for fiber reinforcement of the multiply-connected phase, and (with this in mind) methods have been developed which encourage mechanically strong rod-shaped phases to grow with preferred orientation within mechanically weaker phases. Before discussing the bases for these methods, the condition for preferred growth of the rod-like morphology will be examined.[14]

Consider in Figure 2.19 the relative proportions of the phases in a eutectic mixture for, respectively, plate-like and rod-like growth of one phase in the other.

[13]A. Kelly and W. R. Tyson. "Fiber Strengthened Materials," in *High Strength Materials*, ed. V. F. Zackay, John Wiley and Sons, New York, pp. 578–602 (1965).
[14]F. C. Frank and K. E. Puttick. "Cementite Morphology in Pearlite," *Acta. Met.*, 4:206–210 (1956).

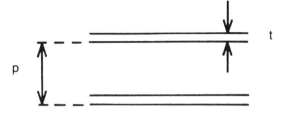

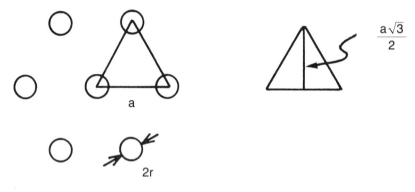

Figure 2.19.

With plate-like growth, the volume fraction of the plate-like phase is

$$\eta = \frac{t\,A}{p\,A} \tag{2.23}$$

where t is the average plate thickness and p is the average separation between plates. A is the interfacial area between the phases.

By considering the unit cell with triangular base, it is evident that, with rod-like growth, the volume fraction of the rod-like phase is

$$\eta = \frac{\left(\dfrac{\pi r^2}{2}\right) l}{\left(\dfrac{a}{2}\right)\left(a\,\dfrac{\sqrt{3}}{2}\right) l} \tag{2.24}$$

where r is the average rod diameter and a is the average rod spacing. l is the average rod length.

If γ is the specific interfacial free energy, i.e., the interfacial free energy per unit area, then the respective interfacial energies per unit volume are

$$\frac{2\gamma A}{pA} = s \text{ for the sheet morphology} \tag{2.25}$$

and

$$\frac{\left(\dfrac{2\pi r\,l}{2}\gamma'\right)}{\left(\dfrac{a}{2}\right)\cdot\left(\dfrac{a\sqrt{3}}{2}\right)l} = s' \text{ for the rod morphology} \tag{2.26}$$

The maximum diffusion distance for redistribution of solute is

$$d = \frac{1}{2}\,p - \frac{1}{2}\,t$$

$$= \frac{1}{2}\,p\,(1 - \eta) \text{ for the sheet morphology} \tag{2.27}$$

and

$$d' = \left(\frac{2}{3}\,\frac{a\sqrt{3}}{2}\right) - r$$

$$= \frac{a}{\sqrt{3}} - a\left(\frac{\eta\sqrt{3}}{2\pi}\right)^{1/2}$$

$$= a\left\{\frac{1}{\sqrt{3}} - \left(\frac{\eta\sqrt{3}}{2\pi}\right)^{1/2}\right\} \text{ for the rod morphology} \tag{2.28}$$

Assuming that $s = s'$, Equations (2.25) and (2.26) give

$$\frac{2\gamma}{p} = \pi r\gamma'\left(\frac{4}{a^2\sqrt{3}}\right) \tag{2.29}$$

Substituting $r = [(\eta a^2\sqrt{3})/(2\pi)]^{1/2}$ from Equation (2.24) into Equation (2.29), we find

$$\frac{a}{p} = \left(\frac{2\pi\eta}{\sqrt{3}}\right)^{1/2}\frac{\gamma'}{\gamma} \tag{2.30}$$

Hence

$$\frac{d'}{d} = \frac{a\left\{ \dfrac{1}{\sqrt{3}} - \left(\dfrac{\eta\sqrt{3}}{2\pi} \right)^{1/2} \right\} \left(\dfrac{2\pi\eta}{\sqrt{3}} \right)^{1/2} \dfrac{\gamma'}{\gamma}}{\dfrac{1}{2} a(1 - \eta)}$$

$$= 2\left\{ \left(\frac{2\pi\eta}{3\sqrt{3}} \right)^{1/2} - \eta \right\} \frac{\gamma'}{(1 - \eta)\gamma}$$

$$= 2(1.105\sqrt{\eta} - \eta)\frac{\gamma'}{(1 - \eta)\gamma} \tag{2.31}$$

Assuming that $\gamma = \gamma'$, i.e., that interface surface curvature does not affect the specific interfacial energy, $d'/d > 1$ when $\eta < 0.78$, i.e., rods are favored at volume fractions of the rod-like phase of 4/5 or less.

Returning to the practice of directional solidification, current high rate solidification (HRS) processes are developments of the Bridgeman crystal growing technique shown schematically in Figure 2.20. The charge is withdrawn from inside two furnaces. It is usual for the lower furnace to be at a temperature which is more below the melting point than the upper furnace is above the melting point, as shown by the temperature profile in Figure 2.20. The reasons for this are as follows. Heat flows into the charge from the hot furnace and out of the charge in the cold furnace. The continuous lines in Figure 2.20 indicate the curvature of the lines of heat flow. In the solid, isothermals (broken lines in Figure

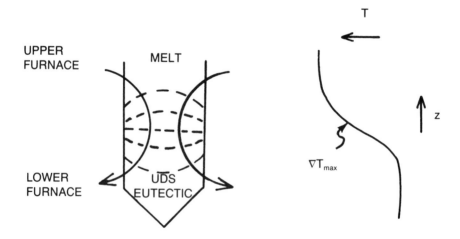

Figure 2.20.

2.18) are orthogonal to the lines of heat flow. Although not strictly true, because of convection processes transferring heat, we shall assume that this is also the case in the liquid. It can be assumed that the usual laws of diffusion (or conduction) of heat apply and that it will be steady state heat flow, i.e.,

$$\nabla^2 T = 0$$

Now, the freezing surface is usually required to coincide with a convex upwards isothermal so that:

1. Convection flow over the surface will be directed so as to sweep impurity inclusions away from the middle of the charge.
2. Sub-boundaries, which propagate normal to the interface at points of strain arising near the crucible wall, will tend to run away from the middle of the solidifying alloy.

Rewriting the steady state heat flow equation as

$$\frac{\partial^2 T}{\partial z^2} = - \frac{\partial^2 T}{\partial x^2} - \frac{\partial^2 T}{\partial y^2}$$

it is evident that convexity in the horizontal xy-plane makes $\partial^2 T/\partial x^2$ and $\partial^2 T/\partial y^2$ both positive. So $\partial^2 T/\partial z^2$ must be negative.

Ideally, it would be best to use the maximum temperature gradient (smallest distance between isothermals in Figure 2.20) that the system can generate, but this occurs across the flat isothermal. So in practice, it is usual to compromise and use a location where the temperature gradient is falling.

2.10 Further examples

1. The melt viscosity typical of thermoplastics (~ 0.1 N m^{-2} s^{-1} = 1 poise) is substantially higher than that typical for formulations used in the manufacture of thermoset prepregs. What problems might this pose when attempting to manufacture thermoplastic prepregs?
2. What materials, what fiber lay-up, and what manufacturing method would you recommend to make hand-held missile launchers that are both portable and disposable?
3. A long cylindrical pressure vessel is to be made using a helical filament winding technique. Calculate the helix angle for the most efficient reinforcement. What factors do you expect will determine the bursting pressure of such a vessel?
4. A γ-ray telescope to be launched from a satellite is housed in a filament wound aramid fiber spherical shell. The sketch in Figure 2.21 shows the spherical winder used to wind the shell. The core onto which the shell is wound is an inflatable elastomeric sphere which can be deflated and removed after curing

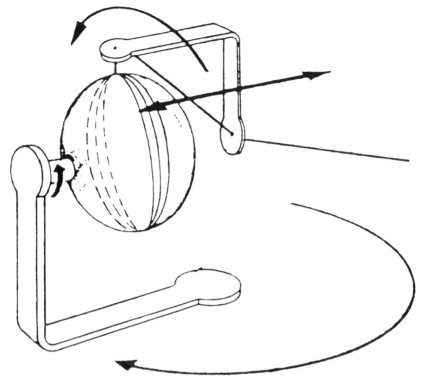

Figure 2.21.

the resin used to bond the fiber. Derive an expression for the pressure that must be maintained inside the inflatable during winding at constant fiber tension.

5. It is proposed to manufacture pultruded aramid fiber/epoxy resin matrix cables for use on off-shore oil rigs. What are the advantages of having the fibers twisted as in a twisted tow, or, for that matter, as in a conventional hemp rope? Describe a manufacturing method that could accomplish such a twist.

6. (Library assignment) Single crystal ceramic whiskers are important candidates for fiber reinforcement of metals. Silicon carbide whiskers, for example, have a specific modulus of $\sim 1.4 \times 10^8 \, m^2 \, s^{-2}$ and strength/weight ratio $\cong 1.2$ km. Summarize the root growth and tip growth models for whisker growth.

7. A certain fiber composite material consists mainly of thin whisker crystals with their crystallographically equivalent long axes in close alignment. As might be expected, its X-ray diffraction pattern resembles the rotation pattern of a single large crystal of the whisker substance rotated about the whisker axis. However, the spots on the diffraction pattern of the fiber composite are

considerably less sharp than those of the single-crystal rotation pattern. It is required to find out in what measure this diffuseness is due to (a) lack of perfectly parallel alignment of the whiskers in the composite, or (b) very small diameter of the individual whisker crystals, or (c) random strains in the whisker crystals. Describe how conditions (a), (b) and (c) would affect the diffraction pattern if each were present singly, and indicate how a quantitative study and measure of them could be made by X-ray diffraction experiments. If all three conditions were present simultaneously, how well do you think you could assess their relative contributions to the observed degree of diffuseness of diffraction spots? You are free to use any X-ray wavelength and any X-ray diffraction technique (University of Bristol Examination, 1970).

8. Investigate the intersections of cylinders produced by radial growth onto fibers when manufacturing ceramic matrix composites by vapor deposition. Consider only cubic arrays of fibers, and assume that the intersections are symmetrical.

3

Anisotropy of stress

"...stress is a field tensor . . . the principal axes of stress do not have to have any special relationship to the symmetry axes of a crystal (laminate) . . ."[15]

3.1 Mathematical representation of second rank tensors

Tensors are classified by their rank. Tensors of rank zero are known as scalars, examples of which are temperature and density. Tensors of rank one are known as vectors, examples of which are electric field at a point and temperature gradient at a point. Vectors can be specified by their components in the directions of a coordinate system of axes. Thus, electric field strength $E = E_1, E_2, E_3$, where the subscripts 1, 2 and 3 denote the orthogonal axes. In this sense, rank two tensors are an extension of vectors. Consider electrical conductivity. In the isotropic case, Figure 3.1(a), the current density j produced by an electrical field E is parallel to E. $j = \sigma E$, where σ is called the electrical conductivity, which implies that j is parallel to E or, that $j_1 = \sigma E_1$, $j_2 = \sigma E_2$ and $j_3 = \sigma E_3$.

In the anisotropic case, Figure 3.1(b), experiment reveals that

$$j_1 = \sigma_{11}E_1 + \sigma_{12}E_2 + \sigma_{13}E_3$$
$$j_2 = \sigma_{21}E_1 + \sigma_{22}E_2 + \sigma_{23}E_3$$
$$j_3 = \sigma_{31}E_1 + \sigma_{32}E_2 + \sigma_{33}E_3$$

What this implies is that, with the field along axis 1 — although E_2 and E_3 are both zero — j_1, j_2 and j_3 all exist. See Figure 3.2.

By measuring j_1, j_2 and j_3, the coefficients can be determined. In general, j will not be parallel to E. By convention, the nine components of conductivity are

---

[15]J. F. Nye. *University of Bristol Lectures* (1965). See also J. F. Nye. *Tensor Properties of Crystals*, Oxford University Press (1957).

61

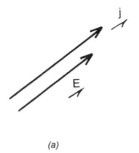

(a)

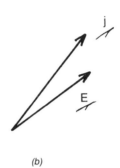

(b)

Figure 3.1.

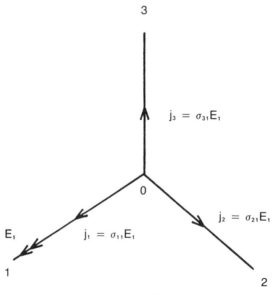

Figure 3.2.

written in matrix form inside square brackets, thus:

$$\begin{bmatrix} \sigma_{11} & \sigma_{12} & \sigma_{13} \\ \sigma_{21} & \sigma_{22} & \sigma_{23} \\ \sigma_{31} & \sigma_{32} & \sigma_{33} \end{bmatrix}$$

This matrix should be regarded as a symbol for conductivity; it symbolizes a second rank tensor.

Thus, a second rank tensor has nine components (cf. a vector which has three components). As will become evident, it is often convenient to consider separately the components on the leading diagonal, that is σ_{11}, σ_{22} and σ_{33}, and the off-diagonal components. Note the use of subscripts. Scalar quantities have no subscript, vectors have one, and second rank tensors have two. That is, the rank of a tensor is indicated by the number of subscripts required to specify it.

Three important second rank tensor properties of composite materials are electrical conductivity (σ) which relates electric field (E) to current density (j), thermal conductivity (k) which relates temperature gradient ($-\operatorname{grad} T$) to heat flow density (h), and stress (σ) which relates unit vector (l) perpendicular to unit area to the force (p) acting across that area.

Electrical conductivity will now be considered further, as representing the general case. In shorthand notation,

$$j_i = \sum_j \sigma_{ij} E_j$$

where i and j can both take the values 1, 2 and 3. j_i is usually shortened further to

$$j_i = \sigma_{ij} E_j$$

that is the symbol for summation is omitted in accordance with the convention that, if a subscript occurs twice within an algebraic term, summation is implied. It is necessary to distinguish between two kinds of suffix. i in the above equation is the free suffix, and j is the dummy suffix. Later, we shall use letters other than j to denote dummy suffices. A dummy suffix is recognizable by its location in an algebraic expression, it occurs twice on one side—the right hand side of the above equation. The isotropic case is $j_i = \sigma E_i$.

If a suffix occurs on the left hand side of the equation, it also occurs on the right hand side.

The choice of coordinate axes is arbitrary. Hence the numerical values for σ are also arbitrary. However, the property measured, the electrical conductivity, is the same. This observation leads to the law of transformation of second rank tensor components.

Consider on old axes

$$p_i = T_{ij} q_j$$

By transformation of vector components,

$$p_i' = a_{ij}p_j$$

where $'$ denotes new axes and a_{ij} are the direction cosines between old and new axes.

Since

$$p_j = T_{jk}q_k,$$

$$p_i' = a_{ij}T_{jk}q_k$$

By transformation of vector components,

$$q_k = a_{lk}q_l'$$

Hence

$$p_i' = a_{ij}T_{jk}a_{lk}q_l'$$

$$= a_{ij}a_{lk}T_{jk}q_l'$$

$$= T_{il}'q_l'$$

So

$$T_{il}' = a_{ij}a_{lk}T_{jk}$$

Thus

$$T_{ij}' = a_{ik}a_{jl}T_{kl} \tag{3.1}$$

is the relationship between the T's.

Expand Equation (3.1) in two parts. k is a dummy suffix, implying summation over k.

Therefore

$$T_{ij}' = a_{i1}a_{jl}T_{1l} + a_{i2}a_{jl}T_{2l} + a_{i3}a_{jl}T_{3l}$$

Since each of the direction cosines a_{jl} can be expanded as the sum of three terms there are nine terms altogether.

For the reverse transformation, from new to old axes, the relationship between the T's is

$$T_{ij} = a_{ki}a_{lj}T_{kl}' \tag{3.2}$$

Summarizing the transformation laws, for scalars $\phi' = \phi$, for vectors $p_i' = a_{ij}p_j$, and for second rank tensors $T_{ij}' = a_{ik}a_{jl}T_{kl}$.

3.2 Homogeneous stress

There are two kinds of forces in nature—body forces, such as that due to gravity, and surface forces, such as stress forces. Stress forces are transmitted across a surface such that the material on one side of the surface affects the material on the other side.

The components of stress are defined as indicated in Figure 3.3. An imaginary cube with axes parallel to the reference axes is cut from the solid and the forces represented by arrows in Figure 3.3 are the forces acting on the material inside the cube due to the material outside it. If the cube has unit faces, the forces will be numerically identical to the components of stress shown in the figure. If the directions of the arrows are taken as positive, tensile forces come out positive, and thus automatically determine the sign convention for the shear forces. Looking along axis 1, the forces acting in the 2 and 3 directions are as shown in Figure 3.4.

To maintain rotational stability, $\sigma_{23} = \sigma_{32}$. Hence the σ_{ij} array of nine stress components associated with the three coordinate axes is symmetrical. It can be shown that the σ_{ij} satisfy the transformation laws for second rank tensors. Since the stress tensor is symmetrical, it will have principal axes, i.e., appropriate

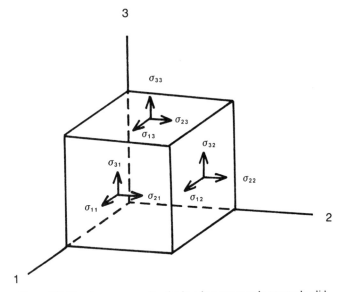

Figure 3.3. The forces on a unit cube in a homogeneously stressed solid.

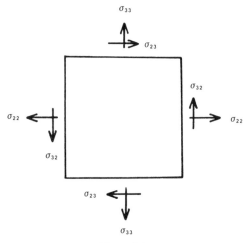

Figure 3.4.

transformation yields

$$
\begin{bmatrix}
\sigma_{11} & \sigma_{12} & \sigma_{13} \\
\sigma_{21} & \sigma_{22} & \sigma_{23} \\
\sigma_{31} & \sigma_{32} & \sigma_{33}
\end{bmatrix}
\rightarrow
\begin{bmatrix}
\sigma_1 & 0 & 0 \\
0 & \sigma_2 & 0 \\
0 & 0 & \sigma_3
\end{bmatrix}
$$

So, although the imaginary unit cube will generally have both normal and shear stresses acting across its faces, there will be one setting of the cube for which there is no shear and the only forces acting on it are normal forces. These normal forces will be of magnitude σ_1, σ_2, σ_3. This result gives us the principal axes of stress, and the principal stresses. Note that, when referred to principal axes it is usual to drop the two suffix notation.

Note also, that there is an important difference between stress and other tensor properties in that the principal axes of stress do *not* have to have any special relationship to the composite or laminate symmetry axes. Stress is like a field that is imposed on the solid from the outside. There are in nature, "field" tensors (such as stress, and electric fields) and "body" tensors which do relate to the symmetry of the solid.

The components of stress can be transformed using the transformation laws for second rank tensors. With fiber reinforced laminates, it usually happens that we have to rotate about a principal axis and, in this case, the transformation law is very simple. Referring to Figure 3.5, let principal axis Ox_3 be the rotation axis. Principal axes Ox_1 and Ox_2 are in the plane of the diagram. Rotate to $1'$, $2'$ and $3'$ ($= 3$) through the angle θ. The matrix of direction cosines

$$
|a_{ij}| =
\begin{vmatrix}
a_{11} & a_{12} & a_{13} \\
a_{21} & a_{22} & a_{23} \\
a_{31} & a_{32} & a_{33}
\end{vmatrix}
$$

is

		old axes		
		x_1	x_2	x_3
	x_1'	$\cos\theta$	$\sin\theta$	0
new axes	x_2'	$-\sin\theta$	$\cos\theta$	0
	x_3'	0	0	1

That is

$$x_1' = x_1 \cos\theta + x_2 \sin\theta$$

$$x_2' = -x_1 \sin\theta + x_2 \cos\theta$$

$$x_3' = x_3$$

Suppose that Ox_1, Ox_2 are also principal axes. Then, on old axes,

$$\sigma_{ij} = \begin{bmatrix} \sigma_1 & 0 & 0 \\ 0 & \sigma_2 & 0 \\ 0 & 0 & \sigma_3 \end{bmatrix}$$

We want to know the components of stress on new axes. Using the transformation

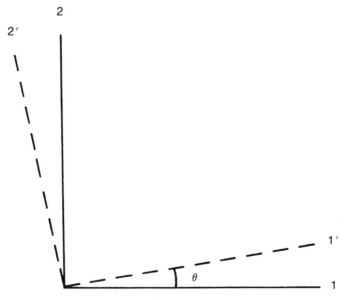

Figure 3.5.

law, these are

$$\sigma'_{ij} = a_{il} a_{jm} \sigma_{lm}$$

$$= a_{i1} a_{j1} \sigma_{11} + a_{i1} a_{j2} \sigma_{12} + a_{i1} a_{j3} \sigma_{13}$$

$$+ a_{i2} a_{j1} \sigma_{21} + a_{i2} a_{j2} \sigma_{22} + a_{i2} a_{j3} \sigma_{23}$$

$$+ a_{i3} a_{j1} \sigma_{31} + a_{i3} a_{j2} \sigma_{32} + a_{i3} a_{j3} \sigma_{33}$$

However, the old axes are principal axes, that is when $l \neq m$, $\sigma_{lm} = 0$. Hence

$$\sigma'_{ij} = a_{i1} a_{j1} \sigma_{11} + a_{i2} a_{j2} \sigma_{22} + a_{i3} a_{j3} \sigma_{33}$$

and

$$\sigma'_{11} = a_{11} a_{11} \sigma_{11} + a_{12} a_{12} \sigma_{22} + a_{13} a_{13} \sigma_{33}$$

$$\sigma'_{12} = a_{11} a_{21} \sigma_{11} + a_{12} a_{22} \sigma_{22} + a_{13} a_{23} \sigma_{33}$$

$$\sigma'_{13} = a_{11} a_{31} \sigma_{11} + a_{12} a_{32} \sigma_{22} + a_{13} a_{33} \sigma_{33}$$

$$\sigma'_{22} = a_{21} a_{21} \sigma_{11} + a_{22} a_{22} \sigma_{22} + a_{23} a_{23} \sigma_{33}$$

$$\sigma'_{23} = a_{21} a_{31} \sigma_{11} + a_{22} a_{32} \sigma_{22} + a_{23} a_{33} \sigma_{33}$$

$$\sigma'_{33} = a_{31} a_{31} \sigma_{11} + a_{33} a_{32} \sigma_{22} + a_{33} a_{33} \sigma_{33}$$

Substituting from our table of direction cosines, we end up with

$$\sigma'_{11} = \sigma_1 \cos^2 \theta + \sigma_2 \sin^2 \theta$$

$$\sigma'_{22} = \sigma_1 \sin^2 \theta + \sigma_2 \cos^2 \theta$$

$$\sigma'_{12} = -\sigma_1 \sin \theta \cos \theta + \sigma_2 \sin \theta \cos \theta$$

which, using

$$\cos^2 \theta + \sin^2 \theta = 1$$

and

$$\cos^2 \theta - \sin^2 \theta = \cos 2\theta$$

can be rewritten as

$$\sigma'_{11} = \frac{1}{2}(\sigma_1 + \sigma_2) - \frac{1}{2}(\sigma_2 - \sigma_1)\cos 2\theta$$

$$\sigma'_{22} = \frac{1}{2}(\sigma_1 + \sigma_2) + \frac{1}{2}(\sigma_2 - \sigma_1)\cos 2\theta \qquad (3.3)$$

$$\sigma'_{12} = \frac{1}{2}(\sigma_2 - \sigma_1)\sin 2\theta$$

None of the other components of stress change from their values on the old axes.

3.3 Worked example

In the coordinate system x_1, x_2, x_3, the state of stress in a fiber reinforced laminate is

$$\begin{bmatrix} 7.2 & 0 & 0 \\ 0 & 1.5 & 0 \\ 0 & 0 & 0.8 \end{bmatrix} \text{ MN m}^{-2}$$

What does this statement mean? Calculate the components of stress referred to a coordinate system of axes rotated with respect to the first set by an angle of $45°$ about x_3.

First part: the principal stresses are 7.2 MN m^{-2}, 1.5 MN m^{-2}, 0.8 MN m^{-2}.
Second part:

$$\sigma'_{ij} = \begin{bmatrix} \sigma_1\cos^2\theta + \sigma_2\sin^2\theta & -\sigma_1\cos\theta\sin\theta + \sigma_2\sin\theta\cos\theta & 0 \\ & \sigma_1\sin^2\theta + \sigma_2\cos^2\theta & 0 \\ & & \sigma_3 \end{bmatrix}$$

$\theta = 45°$, so $\sin\theta = \cos\theta = \dfrac{1}{\sqrt{2}}$

Therefore

$$\sigma'_{ij} = \begin{bmatrix} 4.35 & -2.85 & 0 \\ & 4.35 & 0 \\ & & 0.8 \end{bmatrix} \text{ MN m}^{-2}$$

3.4 Mohr circle

The Mohr circle, Figure 3.6 is a geometrical construction of Equations (3.3), that is, it shows how the transformed components of stress change during rotation about a principal axis.

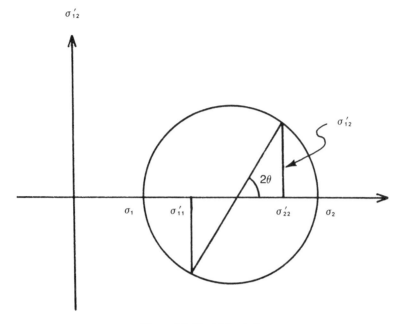

Figure 3.6. The Mohr circle.

The coordinates represent the components of normal and shear stress respectively. Inspection of the Mohr circle construction enables Equations (3.3) to be written down. It also shows that the shear stress is zero along the principal axes, and is a maximum when $2\theta = \pi/2$, i.e., when $\theta = \pi/4$ ("the 45° position"). The maximum value of the shear stress, $\sigma'_{12} = \frac{1}{2}(\sigma_2 - \sigma_1)$, is the radius of the circle. During rotation, the normal components of stress stay inside the range defined by the principal stresses, i.e., the principal stresses are the extreme values of the normal stresses.

The center of the circle stays put, i.e., $\frac{1}{2}(\sigma'_{11} + \sigma'_{22})$ is invariant and equals $\frac{1}{2}(\sigma_1 + \sigma_2)$. Finally, if σ'_{11}, σ'_{22}, σ'_{12} are known, the Mohr circle can be used to calculate the principal stresses σ_1, σ_2.

3.5 Geometrical representation of second rank tensors[15]

Consider the quadric

$$S_{ij}x_ix_j = 1 \tag{3.3}$$

where $S_{ij} = S_{ji}$. Both suffices occur on one side only so both are dummy suffices.

Expanding, we get

$$S_{11}x_1^2 + S_{22}x_2^2 + S_{33}x_3^2 + 2S_{12}x_1x_2 + 2S_{23}x_2x_3 + 2S_{31}x_3x_1 = 1 \qquad (3.4)$$

Notice the quadric form of this equation; it is the equation to a quadric, i.e., an ellipsoid or hyperboloid, with center at the origin.

Transferring the x's to new axes, we get

$$x_i = a_{ki}x_k'$$

and

$$x_j = a_{lj}x_l'$$

Hence

$$S_{ij}a_{ki}x_k'a_{lj}x_l' = 1$$

The original equation can be rewritten as $S_{kl}'x_k'x_l' = 1$, so

$$S_{kl}' = a_{ki}a_{lj}S_{ij}$$

which is the same as the second rank tensor transformation law, i.e., the coefficients S_{ij} transform algebraically in an identical way to second rank tensor components.

Hence, we can exploit the geometrical properties of quadrics in order to represent second rank tensors. In particular, a quadric possesses three principal axes, and so does a second rank tensor.

Transforming Equation (3.4) to principal axes gives

$$S_1x_1^2 + S_2x_2^2 + S_3x_3^2 = 1 \qquad (3.5)$$

which—depending on whether S_1, S_2, S_3 are positive or negative—is the equation of either an ellipsoid or a hyperboloid. Similarly for a second rank tensor, we have

$$\begin{bmatrix} S_{11} & S_{12} & S_{13} \\ S_{12} & S_{22} & S_{23} \\ S_{13} & S_{23} & S_{33} \end{bmatrix} \rightarrow \begin{bmatrix} S_1 & 0 & 0 \\ 0 & S_2 & 0 \\ 0 & 0 & S_3 \end{bmatrix}$$

In response to the question, what are the surfaces which represent this tensor, if S_1, S_2, S_3 are all positive, the surface is an ellipsoid with semi-axis lengths $1/\sqrt{S_1}$, $1/\sqrt{S_2}$, $1/\sqrt{S_3}$, respectively, see Figure 3.7 (a).

If one or more of the S's is negative, the corresponding semi-axis length

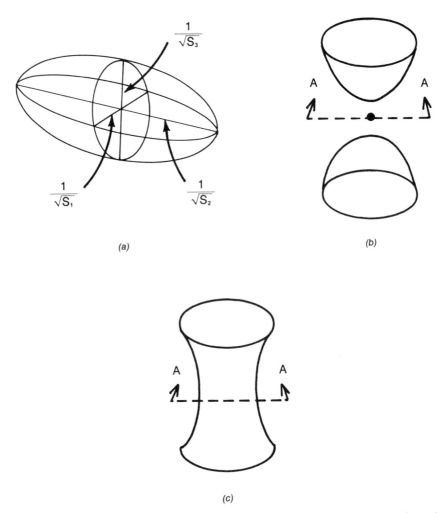

$\dfrac{1}{\sqrt{S_3}}$

$\dfrac{1}{\sqrt{S_1}}$ $\dfrac{1}{\sqrt{S_2}}$

A   A

(a)

(b)

A   A

(c)

Figure 3.7. Representation quadrics for second rank tensors. (a) All three principal properties positive, (b) two negative and one positive (section A-A is an imaginary ellipse, and the other two sections, lying in the plane of the paper are hyperbolas), (c) two positive and one negative (section A-A is a real ellipse). The case of all three principal properties negative gives an imaginary ellipsoid.

becomes imaginary and this gives rise to the two kinds of hyperboloid sketched in Figure 3.7(b) and (c).

Note that on principal axes, and only on principal axes, the relationships $p_i = S_{ij}q_j$ between the two vectors p_i and q_j yield the simple proportionality relationships $p_1 = S_1q_1$, $p_2 = S_2q_2$, $p_3 = S_3q_3$. The orthogonal property, i.e., that the directions of the maximum, minimum and mini-maximum values of any symmetric second rank tensor property are at right angles to each other, is a remarkable property.

3.6 Worked examples

1. In a composite consisting of three sets of different gage carbon fibers set in an insulating matrix—the sets obliquely oriented with respect to each other—the

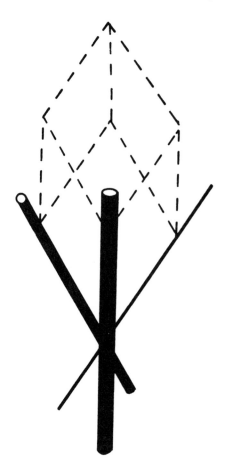

Figure 3.8. Three different gage wires obliquely oriented with respect to each other.

directions of maximum, minimum and mini-maximum conductivities are at right angles to each other. Explain.

The fiber orientations are shown schematically in Figure 3.8. Since conductivity (electrical and thermal) is a second rank tensor property, there will be three mutually orthogonal axes, the principal axes, along which the overall conductivity of the composite will have maximum, minimum and mini-maximum values respectively. This is true, whatever the relative gages, whatever the relative volume fractions and whatever the relative orientations of the three sets of wires. The maximum, minimum and mini-maximum conductivities are known as the principal conductivities, the existence of which follows from Onsager's principle.

2. The thermal conductivity of a fiber reinforced composite is given by

$$k_{ij} = \begin{bmatrix} 0.50 & 0 & 0 \\ 0 & 0.63 & 0 \\ 0 & 0 & 0.34 \end{bmatrix} \text{ W m}^{-1} \text{ K}^{-1}$$

What does this statement mean?

If the temperature gradient is given by the components

$$\begin{bmatrix} -1 \\ 0 \\ -1 \end{bmatrix} \text{K m}^{-1}$$

in what direction is the maximum rate of decrease of temperature? Find the flow of heat. In what direction is the heat flowing? Sketch an appropriate section of the representation quadric to illustrate your answer (University of Bristol 1967).

As stated, the k_{ij} describe a symmetric second rank tensor referred to principal axes.

In an isotropic solid, heat conduction obeys the law

$$\frac{\partial h}{\partial t} = -kA \frac{\partial T}{\partial x}$$

or, since h is a vector quantity

$$\frac{\partial h_i}{\partial t} = -k \frac{\partial T}{\partial x_j}$$

per unit area where t denotes time and T temperature.

In an anisotropic solid $\partial h_i / \partial t$ is generally not parallel to $\partial T / \partial x_j$. So,

$$\frac{\partial h_i}{\partial t} = -k_{ij} \frac{\partial T}{\partial x_j}$$

Here,

$$\frac{\partial h_1}{\partial t} = -0.5 \frac{\partial T}{\partial x_1}$$

$$\frac{\partial h_2}{\partial t} = -0.63 \frac{\partial T}{\partial x_2}$$

$$\frac{\partial h_3}{\partial t} = -0.34 \frac{\partial T}{\partial x_3}$$

The direction cosines of the maximum $\partial T/\partial x_j$ are $(1/\sqrt{2})\,(-1,\,0,\,-1)$. Hence the heat flow is

$$\frac{\partial h_i}{\partial t} = \begin{bmatrix} 0.5 & 0 & 0 \\ 0 & 0.63 & 0 \\ 0 & 0 & 0.34 \end{bmatrix} \begin{bmatrix} -1 \\ 0 \\ -1 \end{bmatrix} \text{J m}^{-2}\text{ s}^{-1}$$

$$= \begin{bmatrix} -0.5 \\ 0 \\ -0.34 \end{bmatrix} \text{J m}^{-2}\text{ s}^{-1}$$

and the direction cosines for the heat flow are

$$\frac{1}{\sqrt{(-0.50)^2 + (-0.34)^2}}\,(-0.50,\,0,\,-0.34)$$

$$= \frac{1}{\sqrt{0.37}}\,(-0.50,\,0,\,-0.34)$$

$$= \frac{6}{6\sqrt{0.37}}\,(-0.50,\,0,\,-0.34)$$

$$= \frac{1}{\sqrt{13}}\,(-3,\,0,\,-2)$$

That is, the heat is flowing in the $[\bar{3}\ 0\ \bar{2}]$ direction.
The representation quadric for thermal conductivity is

$$k_{ij}x_i x_j = 1$$

or, referred to principal axes,

$$k_1 x_1^2 + k_2 x_2^2 + k_3 x_3^2 = 1$$

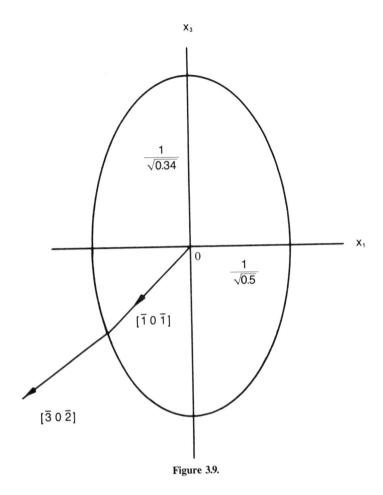

Figure 3.9.

which, for the case considered here, is

$$0.5x_1^2 + 0.63x_2^2 + 0.34x_3^2 = 1$$

This is the equation to an ellipsoid, the $Ox_1 Ox_3$ section of which is sketched in Figure 3.9 together with arrows indicating the $[\bar{1}\,0\,\bar{1}]$ direction of the temperature gradient, and the $[\bar{3}\,0\,\bar{2}]$ direction of the heat flow.

3.7 Neumann's principle[16]

F. Neumann propounded a fundamental principle in regard to the physical properties of crystals. In the context of continuous fiber reinforced composites

[16]F. Neumann. *Vorlesungen über die Theorie der Elastizität*, Leipzig (1885).

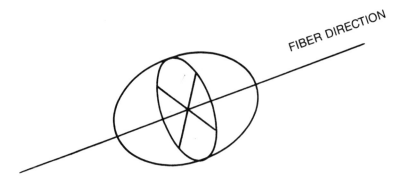

Figure 3.10. Representation quadric for a uniaxial composite.

and laminates, it may be stated as follows. The symmetry of any property must be at least that of the composite.

Consider a three-dimensionally reinforced continuous carbon fiber composite. If the fiber volume fraction is the same in all three dimensions, the characteristic symmetry of the composite is four three-fold axes. However, experiment would reveal that the conductivity, for example is isotropic; the ellipsoidal representation quadric becomes a sphere:

$$\begin{bmatrix} S & 0 & 0 \\ 0 & S & 0 \\ 0 & 0 & S \end{bmatrix}$$

That is, the conductivity has a higher symmetry than the composite, which fits in with Neumann's principle.

Analytically, rotations of $2\pi/3$ would be made about the four three-fold axes and the tensor components caused to remain the same. It can be shown that, for subsequent rotation about any direction by any arbitrary angle, the magnitude of any tensor property would then remain the same.

For uniaxial composites, there exists one unique axis and rotational symmetry about this axis. The fiber direction coincides with the direction of principal conductivity and the representation quadric has rotational symmetry about this axis, see Figure 3.10.

For a single ply (orthorhombic symmetry — see Chapter 5), we have three two-fold axes at right angles to each other. These have to be the principal axes of the conductivity representation quadric.

3.8 Magnitude of a property

Engineers often speak of the conductivity (σ) measured in the fiber direction. In general, the electric field vector $\underline{E}$ and the current density $\underline{J}$ lie in different directions so, what is usually meant by the above statement is that E is applied along the fiber direction, the component of J in that direction is measured and the latter is divided by the former to get the conductivity

$$\sigma = \frac{J^{\shortparallel}}{E} \tag{3.6}$$

Suppose the field is applied in the direction defined by the unit vector l_i

$$l_i = l_1, l_2, l_3$$

The field in this direction

$$E = [El_1, El_2, El_3]$$

The result is simplest when we take the principal axes as reference axes. Then the current density

$$J = [\sigma_1 El_1, \sigma_2 El_2, \sigma_3 El_3]$$

To get $J^{\shortparallel}$, we have to resolve each component along that direction, and add. So

$$J^{\shortparallel} = [l_1^2\sigma_1 + l_2^2\sigma_2 + l_3^2\sigma_3]E \tag{3.7}$$

Comparing Equations (3.6) and (3.7) we see that the magnitude of the conductivity in the direction l_i is

$$\sigma = l_1^2\sigma_1 + l_2^2\sigma_2 + l_3^2\sigma_3$$

and this is a quadric.

On general axes, we have

$$E = El_i$$

$$J^{\shortparallel} = \frac{J \cdot E}{|E|}$$

$$= \frac{J_i E_i}{|E|}$$

in suffix notation, where the suffices imply summation.

Hence, from Equation (3.6) we get for the magnitude of the conductivity in direction l_i,

$$\sigma = \frac{J_i E_i}{|E|^2}$$

$$= \frac{\sigma_{ij} E_j E_i}{|E|^2}$$

$$= \sigma_{ij} l_i l_j \qquad (3.8)$$

since $E_j / |E| = l_j$ and $E_i / |E| = l_i$.

3.9 Worked example

The stress tensor in a laminate, referred to the laminate axes Ox_1, Ox_2, Ox_3 is

$$\sigma_{ij} = \begin{bmatrix} 9 & -5 & 5 \\ -5 & 4 & 3 \\ 5 & 3 & 8 \end{bmatrix} \times \text{kN m}^{-2}$$

Find the components, referred to the same axes, of the normal force N_i and the tangential force T_i exerted across 200 mm² of the planar surface perpendicular to the vector $V = [1,2,2]$ by the material on the $-V_i$ side upon that on the $+V_i$ side (from University of Bristol examination, 1966).

The force acting across unit area is

$$p_i = \sigma_{ij} l_j$$

where l_j is a unit vector perpendicular to that area.

The magnitude of the stress acting in the direction l_i is

$$\sigma = \sigma_{ij} l_i l_j$$

$$= (\sigma_{1j} l_1 + \sigma_{2j} l_2 + \sigma_{3j} l_3) l_j$$

$$= \sigma_{11} l_1^2 + \sigma_{22} l_2^2 + \sigma_{33} l_3^2 + 2(\sigma_{12} l_1 l_2 + \sigma_{13} l_1 l_3 + \sigma_{23} l_2 l_3)$$

Substituting

$$l_1 = \frac{1}{\sqrt{(1)^2 + (2)^2 + (2)^2}} = \frac{1}{3}, \quad l_2 = \frac{2}{3}, \quad l_3 = \frac{2}{3},$$

we find

$$\sigma = \left[9 \cdot \frac{1}{9} + \left\{ 4 \cdot \frac{4}{9} + 8 \cdot \frac{4}{9} \right\} + 2 \left\{ - 5 \cdot \frac{2}{9} + 5 \cdot \frac{2}{9} + 3 \cdot \frac{4}{9} \right\} \right] \times 10^3 \, \text{N m}^{-2}$$

$$= \left[1 + \frac{48}{9} + \frac{8}{3} \right] \times 10^3 \, \text{N m}^{-2}$$

$$= 9 \times 10^3 \, \text{N m}^{-2}$$

By convention, σ is the stress exerted on the material inside a unit cube by the material outside it.

$$\text{Hence } N_i = -9 \times 10^3 (\text{N/m}^2) \times 2 \times 10^{-4} (\text{m}^2)$$
$$= -1.8 \, \text{N}.$$

Since T_i and N_i act at right angles to each other, the dot product

$$[l_1 \ l_2 \ l_3] \cdot [l_1' \ l_2' \ l_3'] = 0$$

where $[l_1' \ l_2' \ l_3']$ is the direction of T_i.
  That is,

$$[1 \ 2 \ 2] \cdot [l_1' \ l_2' \ l_3'] = 0$$

Inspection of a standard stereographic projection for a cubic crystal, or application of the Weiss zone law, reveals that possible indices for $[l_1' \ l_2' \ l_3']$, include $[2 \ \bar{2} \ 1]$, $[0 \ \bar{1} \ 1]$, $[\bar{2} \ \bar{1} \ 2]$ and $[\bar{2} \ 0 \ 1]$. By trial and error, it is found that the shear stress is a maximum for $[2 \ \bar{2} \ 1]$.
  Substituting $l_1' = 2/3$, $l_2' = -2/3$, $l_3' = 1/3$ gives, for the shear stress acting along $[2 \ \bar{2} \ 1]$,

$$\sigma =$$

$$\left[9 \cdot \frac{4}{9} + \left\{ 4 \cdot \frac{4}{9} + 8 \cdot \frac{1}{9} \right\} + 2 \left\{ - 5 \cdot \frac{-4}{9} + 5 \cdot \frac{2}{9} + 3 \cdot \frac{-2}{9} \right\} \right]$$
$$\times 10^3 \, \text{N m}^{-2}$$

$$= \left[4 + \frac{24}{9} + \frac{48}{9} \right] \times 10^3 \, \text{N m}^{-2}$$

$$= 12 \times 10^3 \, \text{N m}^{-2}.$$

Hence

$$T_i = -12 \times 10^3 (\text{N/m}^2) \times 2 \times 10^{-4} (\text{m}^2)$$

$$= -2 \cdot 4 \, \text{N}$$

3.10 Radius normal property

One graphical representation of the variation with orientation of the magnitude of a property is to plot a three-dimensional polar diagram. A more convenient representation is to draw the surface such that

$$r = \frac{1}{\sqrt{\sigma}}$$

This expression gives the representation quadric, Figure 3.11. The quadric is

$$\sigma_{ij} X_i X_j = 1$$

and, at any point P, the coordinates are

$$P = [X_i]$$

$$X_i = r l_i$$

where l_i are the direction cosines for OP.
Hence

$$\sigma_{ij} l_i l_j r^2 = 1$$

$$\sigma_{ij} l_i l_j = \sigma,$$

the magnitude of the property in direction l_i [refer to Equation (3.8)].

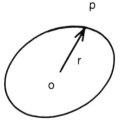

Figure 3.11.

So

$$\sigma r^2 = 1$$

and

$$r = \frac{1}{\sqrt{\sigma}}$$

The semi-axis lengths of the representation quadric are, of course,

$$\frac{1}{\sqrt{\text{principal } \sigma\text{'s}}}$$

This turns out to be a very useful construction since it shows how the magnitudes of properties vary with direction. The semi-principal axis lengths of the representation quadric give us the magnitudes of the maximum, minimum and mini-maximum values of a property. They do not, however, say anything about the angle between E and J. This information is contained in the radius normal. We will first state the so-called radius normal property of the representation quadric, and then justify it.

In Figure 3.12, draw a tangent plane through the intersection of E with the quadric. J is normal to the tangent plane.

This fits in with what we know about the principal axes because, along these directions, E and J coincide and the tangent plane is normal to both.

$$J = (\sigma_1 l_1 E, \ \sigma_2 l_2 E, \ \sigma_3 l_3 E)$$

see Section 3.5. Hence, the direction cosines of J are proportional to

$$\sigma_1 l_1, \ \sigma_2 l_2, \ \sigma_3 l_3$$

P is a point on the representation quadric

$$\sigma_1 x_1^2 + \sigma_2 x_2^2 + \sigma_3 x_3^2 = 1 \tag{3.5}$$

such that OP is parallel to E.

$$P = (r l_1, \ r l_2, \ r l_3)$$

so, for the tangent plane, we have the equation

$$r l_1 \sigma_1 x_1 + r l_2 \sigma_2 x_2 + r l_3 \sigma_3 x_3 = 1$$

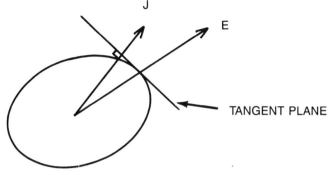

Figure 3.12.

which means that

$$l_1\sigma_1, \ l_2\sigma_2, \ l_3\sigma_3$$

are the direction cosines of the normal to the tangent plane. They are also the direction cosines of J.

Summarizing, if $p_i = S_{ij}q_j$ and $S_{ij} = S_{ji}$, a radius of length r drawn in the representation quadric $S_{ij}x_ix_j = 1$ parallel to the vector field (q) that is applied, produces a vector response $p = S|q| = (1/r^2)|q|$, the direction of which is given by the radius normal.

3.11 Homogeneous strain

First consider, in Figure 3.13, one-dimensional strain, e.g., consider a string before and after extension. Assume that the origin O is fixed. Let u be the displacement of point x. Consider a neighboring point at $x + \Delta x$. The displacement of this point is $u + \Delta u$ and the strain of element Δx is $e = \Delta u/\Delta x$. In the limit, the strain at a point is $e = du/dx$. That is, strain is a derivative of displacement. For small displacements, $\Delta u = (du/dx)\,\Delta x$.

In three dimensions, consider a point with coordinates x_i. Let u_i, a vector, be the displacement of that point. Consider a neighboring point $x_i + \Delta x_i$. The displacement of this point is $u_i + \Delta u_i$, refer to Figure 3.14, and, for small displacements, we may write,

$$\Delta u_i = \frac{\partial u_i}{\partial x_j}\,\Delta x_j$$

$$= e_{ij}\Delta x_j \tag{3.6}$$

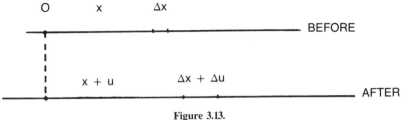

Figure 3.13.

which is a linear relationship between two vectors; the e_{ij} form the components of a second rank tensor, that is they form an array of nine numbers all of which must be specified in order to completely describe the displacement.

Consider now, the special case of Δx_2, Δx_3 both zero. Referring to Figure 3.15

$$\Delta u_1 = e_{11}\Delta x_1 + e_{12}\Delta x_2 + e_{13}\Delta x_3$$

$$= e_{11}\Delta x_1 + 0 + 0$$

$$\Delta u_2 = e_{21}\Delta x_1$$

Hence $e_{11} = \Delta u_1/\Delta x_1$ which is the stretch, i.e., extension per unit length, in direction Ox_1.

$$e_{21} = \frac{\Delta u_2}{\Delta x_1} \cong \frac{\Delta u_2}{\Delta x_1 + \Delta u_1} = \text{an angle}$$

e_{22}, etc. have similar physical interpretations.

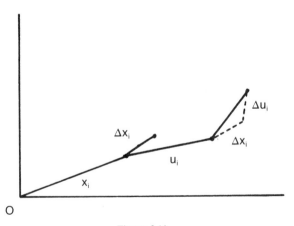

Figure 3.14.

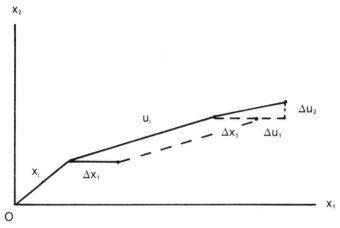

Figure 3.15.

Summarizing, the diagonal components of e_{ij} are stretches, the off-diagonal components are tilts of elements.

For homogeneous strain, the components of e_{ij} are uniform and therefore constants. Hence we can integrate Equation (3.6) to get

$$u_i = e_{ij}x_j \qquad (3.7)$$

All this is for a fixed origin. If the origin moves during deformation, a vector which is a measure of displacement of the origin must be added to Equation (3.7).

Note that e_{ij} is not a strain. It can represent a rotation of the body as well as a strain. This suggests that we should get rid of the rotation in order to obtain a definition of strain. In Figure 3.16 consider a pure rotation. Let the body rotate about an axis through O.

Since the displacement u_i is at right angles to x_i, the scalar product

$$u_i x_i = 0$$

Substituting for u_i from Equation (3.7), we have

$$e_{ij}x_i x_j = 0$$

That is pairs of terms like

$$e_{12}x_1 x_2 + e_{21}x_2 x_1 = 0$$

in which case, all we can say is that

$$e_{12} + e_{21} = 0$$

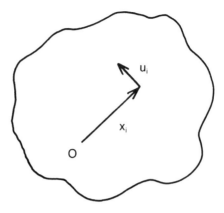

Figure 3.16.

or, generally, that

$$e_{ij} = -e_{ji}$$

which implies that coefficients like e_{11} (the diagonal terms) are zero.

Hence it is evident that e_{ij} is an anti-symmetrical tensor.

Note that, in the above example, the e's are not all zero and yet there is no strain. e is evidently not a satisfactory measure of strain only. We need to subtract the anti-symmetrical part from e and take what is left as our measure of strain.

So, divide e_{ij} into two parts:

$$e_{ij} = \epsilon_{ij} + \omega_{ij}$$

where

$$\left. \begin{array}{l} \epsilon_{ij} = \tfrac{1}{2}(e_{ij} + e_{ji}) \\[2mm] \omega_{ij} = \tfrac{1}{2}(e_{ij} - e_{ji}) \end{array} \right\} \tag{3.8}$$

Since the sum of two tensors is a tensor, ϵ_{ij} is a tensor. Not only that, since $\epsilon_{ij} = \epsilon_{ji}$, ϵ_{ij} is a symmetrical tensor. $\omega_{ij} = -\omega_{ji}$, so ω_{ij} is an anti-symmetrical tensor. ω_{ij} is defined as the rotation. ϵ_{ij} is defined as the strain. Note that $\epsilon_{ij} = 0$ when we have rotation only, as in Figure 3.16. Our definition of rotation is not unambiguous since, in a body which is simultaneously rotating and deforming, there can arise rotation from the deformation as well as from the rotation. The physical meaning of ϵ_{ij} requires explanation. Referring to Figure 3.17, ϵ_{12}, for example, is given by

$$\epsilon_{12} = \tfrac{1}{2}(e_{12} + e_{21})$$

$$= \tfrac{1}{2}\left(\frac{\partial u_1}{\partial x_2} + \frac{\partial u_2}{\partial x_1} \right) \tag{3.9}$$

which is one half the change in angle between a pair of elements originally at right angles to each other.

For diagonal components, ϵ_{11} for example, the ϵ's are the same as the e's.

Now consider a cube. A small deformation changes it to a parallelopiped, see Figure 3.18.

ϵ is a symmetrical second rank tensor, so it possesses principal axes; the components along the principal axes are called the principal strains. The principal axes property of a symmetrical tensor is that, on principal axes, the off-diagonal components vanish, for example

$$\epsilon_{12} = 0$$

Physically, this means that the cube corners in Figure 3.18 are still at right angles. That is, to a first order approximation, if we take a cube defined by the principal

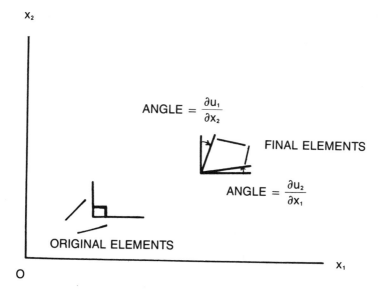

Figure 3.17.

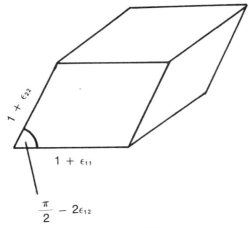

Figure 3.18.

axes, it deforms into a *rectangular* parallelopiped; the edge lengths may become nonequal but corners remain at right angles. Note that this does not mean that the cube does not rotate.

The presence of the factor of 1/2 in Equation (3.9), and of the factor of 2 in the corner angle in Figure 3.18, looks odd. Engineers (as opposed to physicists) leave them out and write

$$\begin{bmatrix} \epsilon_x & \gamma_{xy} & \gamma_{xz} \\ & \epsilon_y & \gamma_{yz} \\ & & \epsilon_z \end{bmatrix}$$

Here, the γ's are the actual changes in the right angles at the corners of our imaginary cube. They are called the engineering strains, or engineering strain components, as opposed to the tensor strain components.

It is important to remember that engineering strain components do not form a tensor, and therefore need to be converted to tensor strain components in order to transform them using tensor transformation laws.

The Mohr circle construction (Section 3.4) can be used for any symmetrical second rank tensor. By way of an example of its uses for transformation of components of strain, consider simple shear, Figure 3.19.

γ is the shear. Engineers call γ the shear strain. The $[e_{ij}]$ array is

$$[e_{ij}] = \begin{bmatrix} 0 & \gamma & 0 \\ 0 & 0 & 0 \\ 0 & 0 & 0 \end{bmatrix}$$

which is unsymmetric.

By definition, Equation (3.8), the strain tensor

$$[\epsilon_{ij}] = \begin{bmatrix} 0 & \tfrac{1}{2}\gamma & 0 \\ \tfrac{1}{2}\gamma & 0 & 0 \\ 0 & 0 & 0 \end{bmatrix} \qquad (3.10)$$

That is, the tensor shear strain is $\tfrac{1}{2}\gamma$. We have inserted the factor of 1/2.

Suppose we wish to know the principal components of strain. That the ϵ_{ij} in Equation (3.10) are not on principal axes can be seen by inspection since the off-diagonal components are not all zero. The Mohr circle construction is sketched in Figure 3.20. The old axes state of strain is represented by the points shown at $\pm\tfrac{1}{2}\gamma$ from the origin on the shear strain axis.

Rotation by 90° on the Mohr circle, i.e., by 45° in real space, is required in order to transform the principal strains. Thus, the Mohr circle tells us that transformation to principal axes gives us

$$[\epsilon_{ij}] = \begin{bmatrix} \tfrac{1}{2}\gamma & 0 & 0 \\ 0 & -\tfrac{1}{2}\gamma & 0 \\ 0 & 0 & 0 \end{bmatrix} \qquad (3.11)$$

The deformation of a square with sides parallel to the principal axes, to a rectangle is an illustration of Equation (3.11). This deformation is sketched in Figure 3.21(a). However, our first order theory tells us that the same change of shape could be accomplished by pure shear at 45° to the principal axes [Equation (3.10)]. This deformation is sketched in Figure 3.21(b).

Notice also that simple shear is equivalent to pure shear plus a rotation, Figure 3.22.

We saw in Section 3.2 that stress is a field tensor, and not a matter tensor. What kind of tensor is strain? Since we can cause strain by the application of stress, it is evident that we can generate strain in any direction by changing the orientation of the stress. However, we can also generate strain by thermal expansion ($\epsilon_{ij} = \alpha_{ij}\Delta T$). This latter strain (matter tensor) is, of course, dependent on the symmetry of a laminate whereas that caused by application of stress (field tensor) is not. The thermal expansion of fiber reinforced laminates is further examined in Chapter 7.

It often happens that a body is simultaneously subjected to temperature change and externally applied stress. In such a case we have a combination of two strain tensors and in general, there will be three sets of principal axes, one for the combined deformation and one each for the strain tensors generated by thermal expansion and application of external stress. An imaginary cube with edges parallel to the principal axes for the overall deformation will deform to a rectangular parallelopiped; the angle between any pair of its edges is still a right angle after the deformation. The same cube subjected only to either the thermal expansion induced strain tensor or the stress induced strain tensor will, because its edges are not parallel to the principal axes of either tensor acting alone, deform to a

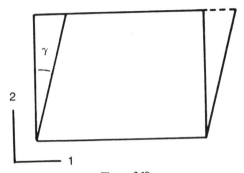

Figure 3.19.

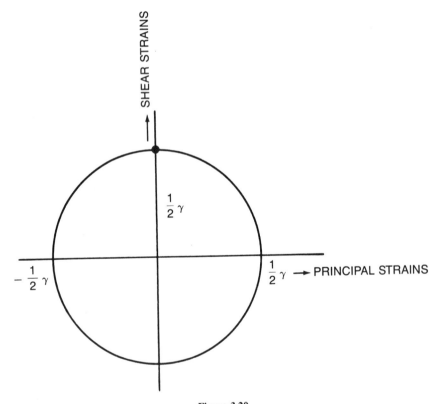

Figure 3.20.

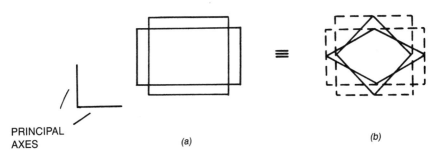

Figure 3.21. Deformation of a square to a rectangle. (a) Sides of square parallel to principal axes. (b) Sides of square at 45° to principal axes.

parallelopiped of general shape and will be further deformed by imposition of the superimposed tensor strain. It is evident, therefore that the state of strain produced by simultaneous action of temperature change and external stress is not identical to that resulting from sequential application of the same temperature change and same external stress.

3.12 Further examples

(1) The state of stress in a laminate is

$$\sigma_{ij} = \begin{bmatrix} 70 & -40 & 0 \\ -40 & 10 & 0 \\ 0 & 0 & 0 \end{bmatrix} \text{MN m}^{-2}$$

Find the components of stress referred to axes rotated clockwise by 30° about axis Ox_1. What are the principal stresses?

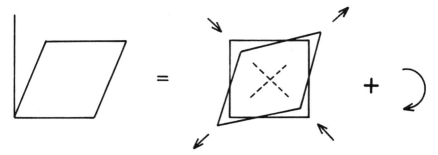

Figure 3.22. Simple shear is pure shear plus a rotation. The broken ±45° lines represent fiber orientations that would resist shear.

(2) Sketch the Mohr circles for (a) uniaxial compression, (b) pure shear stress. For (a) in what direction is the shear stress a maximum and what is its magnitude? For (b), what are the principal stresses and where are the principal axes?

(3) The strain tensor ϵ_{ij} referred to its principal axes has diagonal components 1, 3, 6. Calculate the magnitude of the stretch in the direction which has direction cosines with respect to the principal axes of 0.2, 0.6, 0.775. What is the physical significance of the stretch in the direction with direction cosines $1/\sqrt{3}$, $1/\sqrt{3}$, $1/\sqrt{3}$ (University of Bristol Examination, 1981)?

(4) Find the principal axes of the second rank tensor

$$\begin{bmatrix} 0 & -1 & 0 \\ -1 & -1 & 1 \\ 0 & 1 & 0 \end{bmatrix}$$

If this represents the strain tensor ϵ_{ij}, construct the Ox_1, Ox_2 cross-section of the representation quadric and use it to determine the displacement of the point 3, 2, 0 (University of Bristol Examination, 1982).

(5) A homogeneous two-dimensional strain is specified by the tensor components

$$\epsilon_{11} = \epsilon_{22} = 1.1 \times 10^{-6}; \qquad \epsilon_{12} = 2.3 \times 10^{-6}$$

(a) What is the physical meaning of ϵ_{12}?
(b) What are the magnitudes and directions of the principal strains?
(c) In what direction(s) is the stretch zero?

(6) A_{ij} is an antisymmetric tensor. Find its value in terms of A_{12} etc. for the pair of directions O_i and O_j with direction cosines respectively (2/3, 2/3, 1/3), $(1/\sqrt{2}, -1/\sqrt{2}, 0)$. Interpret this in terms of the components of a vector (University of Bristol Examination, 1972).

(7) Show that, in a material subjected to a small homogeneous strain, the stretch ϵ (extension per unit length) of a line drawn in the direction of the unit vector l_i is given by $\epsilon = \epsilon_{ij} l_i l_j$.

4

Stresses in a plate arising from the presence of a hole

". . .with the highly aeolotropic (anisotropic) spruce the maximum stress rises to 6.37T when the tension (T) is applied parallel to the grain. . ."[17]

4.1 Stresses in a homogeneous isotropic plate due to the presence of an elliptic hole[18]

C. E. Inglis addressed himself to the phenomenon of stress concentration around bolts in the steel plates used to make ships' hulls. The results of his calculation are of fundamental importance, not only to naval architecture but to the general understanding of the importance of stress concentrations in solids.

Many calculations can be simplified by choosing instead of a Cartesian coordinate system, another kind of system that takes advantage of the geometrical symmetry of the problem under consideration. For example, in dislocation theory it is often convenient to use cylindrical coordinates. Similarly, it is convenient to define the displacements near an elliptic* hole in a plate by using elliptic coordinates. These may be introduced as follows:

Figure 4.1 represents a flat homogeneous elastically isotropic slab of uniform thickness, containing an elliptic hole which passes normally through it. The Cartesian axes x, y respectively coincide with the major and minor axes of the elliptic hole, and the z-axis is normal to the plate. Now let $f(x,y) = \alpha$, some constant, be the equation to a curve. If α is allowed to vary we obtain a family of curves. In general one curve of the family will pass through a chosen point, and a neighboring point will in general lie on a neighboring curve of the family, i.e., α is a function of x, y, viz the function denoted by f. If we have two independent families of curves given by the equations

$$f_1(x,y) = \alpha, \text{ a constant: } f_2(x,y) = \beta, \text{ another constant}$$

[17]A. E. Green and G. I. Taylor. "Stress Systems in Aeolotropic Plates III," (1940) published in *Proc. Roy. Soc.*, A 184:181–195 (1945).

[18]C. E. Inglis. "Stresses in a Plate Due to the Presence of Cracks and Sharp Corners," *Trans. Inst. Naval Architects*, 55:219–241, plus plate XXVIII (1913).

*An elliptic hole is a hole whose surface has the form of an elliptic cylinder, i.e., it is not ellipsoidal. The axis of the cylinder is perpendicular to the plate.

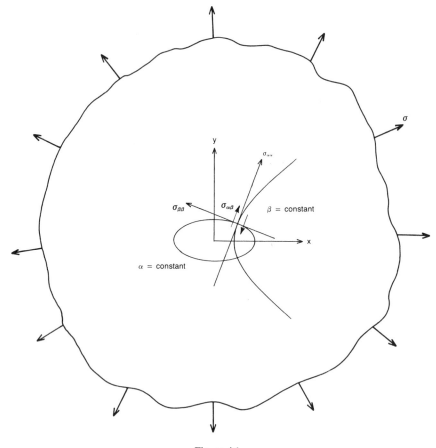

Figure 4.1.

so that, in general, one curve of each family passes through a chosen point, then a point may be determined by the values of α, β which belong to the curves that pass through it, and a neighboring point will be determined by the neighboring values, $\alpha + d\alpha$, $\beta + d\beta$. If the two families of curves cut each other everywhere at right angles, such quantities α, β are called orthogonal curvilinear coordinates of the point.

Suppose the plate in Figure 4.1 is loaded by tractions applied at the outer edge in directions parallel to its surface. At any point well inside the planar surface let u_α, u_β denote the displacements in the directions of the normals to α and β. Let $e_{\alpha\alpha}$, $e_{\beta\beta}$, $e_{\alpha\beta}$ denote the two stretches and the tilt corresponding to these displacements.

$$e_{ij} = \begin{bmatrix} e_{\alpha\alpha} & e_{\alpha\beta} \\ & e_{\beta\beta} \end{bmatrix}$$

(Note that e_{ij} is not a strain. It can represent a rotation or a strain.)
The formulae which follow are derived by A. E. H. Love,[12] pp. 51–55.

$$e_{\alpha\alpha} = h_1 \frac{\delta u_\alpha}{\delta\alpha} + h_1 h_2 u_\beta \frac{\delta}{\delta\beta}\left(\frac{1}{h_1}\right)$$

$$e_{\beta\beta} = h_2 \frac{\delta u_\beta}{\delta\beta} + h_1 h_2 u_\alpha \frac{\delta}{\delta\alpha}\left(\frac{1}{h_2}\right)$$

$$e_{\alpha\beta} = \frac{h_1}{h_2}\cdot\frac{\delta}{\delta\alpha}(h_2 u_\beta) + \frac{h_2}{h_1}\cdot\frac{\delta}{\delta\beta}(h_1 u_\alpha)$$

where

$$h_1^2 = \left(\frac{\delta\alpha}{\delta x}\right)^2 + \left(\frac{\delta\alpha}{\delta y}\right)^2, \qquad h_2^2 = \left(\frac{\delta\beta}{\delta x}\right)^2 + \left(\frac{\delta\beta}{\delta y}\right)^2$$

The sum of the terms on the leading diagonal of the symmetric part of the tensor e_{ij} equals the fractional change in volume, i.e., the dilation (Δ).

$$\Delta = h_1 h_2 \left\{\frac{\delta}{\delta\alpha}\left(\frac{u_\alpha}{h_2}\right) + \frac{\delta}{\delta\beta}\left(\frac{u_\beta}{h_1}\right)\right\}$$

The rotation ω is given by

$$2\omega = h_1 h_2 \left\{\frac{\delta}{\delta\alpha}\left(\frac{u_\beta}{h_2}\right) - \frac{\delta}{\delta\beta}\left(\frac{u_\alpha}{h_1}\right)\right\}$$

For the particular problem under consideration the curvilinear coordinates α, β are such that

$$x = c \cosh \alpha \cos \beta$$
$$y = c \sinh \alpha \sin \beta$$

i.e., $x + iy = c \cosh (\alpha + i\beta)$

c is a constant

α = constant is accordingly the family of ellipses

$$\frac{x^2}{c^2 \cosh^2\alpha} + \frac{y^2}{c^2 \sinh^2\alpha} = 1$$

$$\left(\text{cf. } \frac{x^2}{a^2} + \frac{y^2}{b^2} = 1\right)$$

β = constant is the orthogonal family of hyperbolae

$$\frac{x^2}{c^2 \cos^2\beta} - \frac{y^2}{c^2 \sin^2\beta} = 1$$

$$(\text{cf. } \frac{x^2}{a^2} - \frac{y^2}{b^2} = 1)$$

One ellipse and one hyperbola are sketched in Figure 4.1. In this case, the modulus of transformation h is given by

$$h_1^2 = h_2^2 = h^2 = \frac{2}{c^2(\cosh 2\alpha - \cos 2\beta)}$$

The displacements are given by

$$u = \frac{u_\alpha}{h} = A_n\{(n + p)e^{-(n-1)\alpha}\cos(n + 1)\beta + (n - p)e^{-(n+1)\alpha}\cos(n - 1)\beta\} + \phi$$

$$v = \frac{u_\beta}{h} = A_n\{(n - p)e^{-(n-1)\alpha}\sin(n + 1)\beta + (n + p)e^{-(n+1)\alpha}\sin(n - 1)\beta\} + \psi$$

where $p = 3 - 4\nu$, ν is Poisson's ratio, and ϕ and ψ are conjugate functions of α and β satisfying Laplace's equation. (Laplace's equation is an equation of continuity. In electricity, for example, Laplace's equation is

$$\frac{\partial^2 E}{\partial x^2} + \frac{\partial^2 E}{\partial y^2} + \frac{\partial^2 E}{\partial z^2} = 0$$

and is often written $\nabla^2 E = 0$. E is the electric field and the vector operator ∇ is $\partial/\partial x\ \partial/\partial y\ \partial/\partial z$.) The stress components are

$$\sigma_{\alpha\alpha} = \frac{E}{1 + \nu}\left\{e_{\alpha\alpha} + \frac{\nu}{1 - 2\nu}\Delta\right\}$$

$$\sigma_{\beta\beta} = \frac{E}{1 + \nu}\left\{e_{\beta\beta} + \frac{\nu}{1 - 2\nu}\Delta\right\}$$

$$\sigma_{\alpha\beta} = \frac{E}{2(1 + \nu)}e_{\alpha\beta}$$

where E is Young's modulus.
The suffix convention used here is that component σ_{ij} acts across the surface normal to axis j and in the direction i.

In Figure 4.1 let the elliptic hole be defined by the ellipse $\alpha = \alpha_0$ and let the plate be subjected to a tensile stress σ applied at the outer edge and acting in all directions parallel to the plate (biaxial tension).

Since the forces acting across a free surface must be zero,

$$\sigma_{\alpha\alpha} = 0, \quad \sigma_{\alpha\beta} = 0 \text{ when } \alpha = \alpha_0$$

At a distance from the hole, that is when α is large, $\sigma_{\alpha\alpha} = \sigma_{\beta\beta} = \sigma, \sigma_{\alpha\beta} = 0$. With these boundary conditions, the components of stress at any point are

$$\left. \begin{array}{l} \sigma_{\alpha\alpha} = \dfrac{\sigma \sinh 2\alpha(\cosh 2\alpha - \cosh 2\alpha_0)}{(\cosh 2\alpha - \cos 2\beta)^2} \\[2em] \sigma_{\beta\beta} = \dfrac{\sigma \sinh 2\alpha(\cosh 2\alpha + \cosh 2\alpha_0 - 2\cos 2\beta)}{(\cosh 2\alpha - \cos 2\beta)^2} \\[2em] \sigma_{\alpha\beta} = \dfrac{\sigma \sinh 2\beta(\cosh 2\alpha - \cos 2\beta)}{(\cosh 2\alpha - \cos 2\beta)^2} \end{array} \right\} \quad (4.1)$$

Across the surface of the elliptic hole, that is where $\alpha = \alpha_0$,

$$\sigma_{\beta\beta} = \frac{2\sigma \sinh 2\alpha_0}{\cosh 2\alpha_0 - \cos 2\beta} \quad (4.2)$$

Now consider the case where the plate is subjected to a tensile stress σ acting parallel to the minor axis of the ellipse, see Figure 4.2(a). The boundary conditions are $\sigma_{\alpha\alpha} = \sigma_{\alpha\beta} = 0$ when $\alpha = \alpha_0$ and, since one principal stress equals σ and the other is zero,

$$\sigma_{\alpha\alpha} = \frac{\sigma}{2}(1 - \cos 2\beta)$$

$$\sigma_{\beta\beta} = \frac{\sigma}{2}(1 + \cos 2\beta)$$

$$\sigma_{\alpha\beta} = -\frac{\sigma}{2}\sin 2\beta$$

when α is large.

Of particular interest are the values of $\sigma_{\beta\beta}$ parallel to the surface of the elliptic hole. These are

$$\sigma_{\beta\beta} = \frac{\sigma(\sinh 2\alpha_0 + e^{2\alpha_0}\cos 2\beta - 1)}{\cosh 2\alpha_0 - \cos 2\beta} \quad (4.3)$$

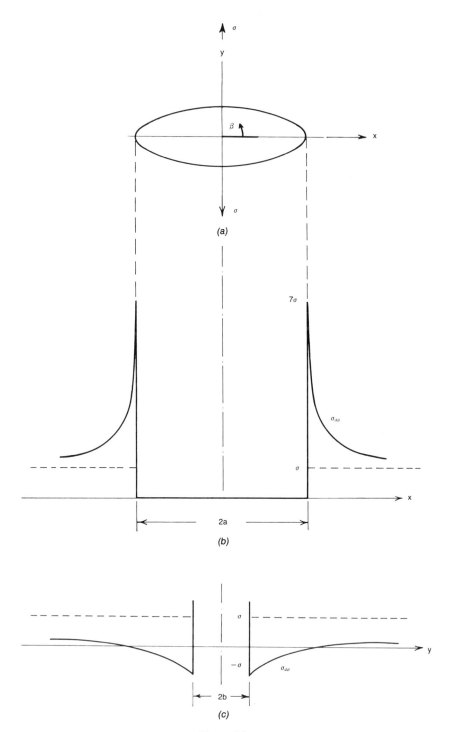

Figure 4.2.

Let the semi-major and semi-minor axes of the ellipse be a and b respectively, that is

$$a = c \cosh \alpha_0 \quad \text{and} \quad b = c \sinh \alpha_0$$

At the ends of the major axis, $\beta = 0$, so

$$\sigma_{\beta\beta} = \sigma \left(1 + \frac{2a}{b} \right) \tag{4.4}$$

and is tensile.

Substituting $\varrho = b^2/a$ for the smaller vertix radius of the ellipse, this becomes

$$\sigma_{\beta\beta} = \sigma \left(1 + 2 \sqrt{\frac{a}{\varrho}} \right) \tag{4.5}$$

In the limit of a very narrow elliptic hole, the hole is geometrically indistinguishable from a crack; refer to Figure 4.3.

Hence it is concluded that the stress concentration at the edge of a crack is

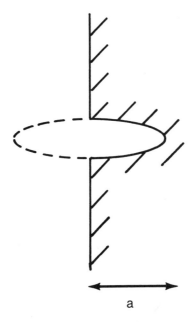

Figure 4.3. Showing the geometrical equivalence of a surface crack (depth a) to an elliptic hole (length $2a$).

given by

$$\frac{\sigma_{concentration}}{\sigma_{distant}} = \left(1 + 2\sqrt{\frac{a}{\varrho}}\right)$$

Typically, surface flaws have depths of the order of 1 μm so, taking an interatomic spacing as order of magnitude for the radius of curvature at the edge of a crack, we have $a/\varrho \cong 1\mu$m$/1$ Å $= 10^4$ and hence a predicted stress amplification of about 100.

At the end of the minor axis, $\beta = \pi/2$ so

$$\sigma_{\beta\beta} = -\sigma$$

and is compressive.

Moving along the x-axis away from the hole, $\sigma_{\beta\beta}$ rapidly decreases from $\sigma(1 + 2a/b)$ at the hole to its normal value σ, as shown in Figure 4.2(b). Advancing along the y-axis, $\sigma_{\beta\beta}$ which is compressive at the hole, soon changes to a small tensile stress which gradually dies out, see Figure 4.2(c).

It is important to note that the stress concentration at the end of the major axis is determined by the ratio a/b, that is, it depends on the shape of the ellipse. For a circular hole, $a = b$ and $\sigma_{\beta\beta} = 3\sigma$, that is the stress is raised locally by a factor of three, and this factor of three concentration is independent of the size of the hole. Anticipating Section 4.2, this is not the case for anisotropic plates; the strength of a fiber reinforced laminate, for example, bears an inverse relationship to the size of any hole drilled in it.

4.2 Stress distribution around a circular hole in an infinite anisotropic plate subjected to uniaxial tension

This problem was solved analytically before World War II by A. E. Green and G. I. Taylor[17] for the case of a plate with two directions of symmetry at right angles to one another. Their method makes use of the Airy stress function χ (see Chapter 9). In the most general case, the equation governing two-dimensional stress distribution in an anisotropic plate is

$$s_{22}\frac{\partial^4\chi}{\partial x^4} - 2s_{26}\frac{\partial^4\chi}{\partial x^2\partial y^2} + (2s_{12} + s_{66})\frac{\partial^4\chi}{\partial x^2\partial y^2} - 2s_{16}\frac{\partial^4\chi}{\partial x^2\partial y^3} + s_{11}\frac{\partial^4\chi}{\partial y^4} = 0$$

$$(4.6)$$

The s_{ij} are components of compliance (see Chapter 5).

By changing Ox and Oy to coincide with the fiber direction and transverse direction, the problem reduces to one of finding a solution, subject to boundary conditions, of the equation

$$\left(\frac{\partial^2}{\partial x^2} + \alpha_1 \frac{\partial^2}{\partial y^2} \right)\left(\frac{\partial^2}{\partial x^2} + \alpha_2 \frac{\partial^2}{\partial y^2} \right) \chi = 0$$

where

$$\alpha_1 \alpha_2 = \frac{s_{11}}{s_{22}}, \quad \alpha_1 + \alpha_2 = \frac{(s_{66} + 2s_{12})}{s_{22}}$$

A solution was first obtained by Green and Taylor using functions of a complex variable. In the case of uniaxial tensile stress (σ) applied to an infinite anisotropic plate containing a circular hole, the circumferential stress around the edge of the hole is given (in Inglis' notation) by

$$\sigma_{\beta\beta} = \sigma \left\{ \frac{(1 + \gamma_1)(1 + \gamma_2)(1 + \gamma_1 + \gamma_2 - \gamma_1\gamma_2 - 2\cos 2\theta)}{(1 + \gamma_1^2 - 2\gamma_1 \cos 2\theta)(1 + \gamma_2^2 - 2\gamma_2 \cos 2\theta)} \right\} \tag{4.7}$$

$\theta = (\pi/2) - \beta$ in Inglis' notation,

$$\gamma_1 = (\alpha_1^{1/2} - 1)/(\alpha_1^{1/2} + 1),$$

$$\gamma_2 = (\alpha_2^{1/2} - 1)/(\alpha_2^{1/2} + 1)$$

Equation (4.7) is to be compared with $\sigma_{\beta\beta} = 3\sigma$ for an isotropic plate. On coordinate axes, the stresses are

$$\sigma_{xx} = \sigma_{\beta\beta} \sin^2 \theta, \quad \sigma_{yy} = \sigma_{\beta\beta} \cos^2 \theta, \quad \sigma_{xy} = -\tfrac{1}{2} \sigma_{\beta\beta} \sin 2\theta$$

(The more general problem of stresses in an anisotropic plate containing an elliptic hole was first solved by C. B. Smith.[19])

In their original publication, Green and Taylor estimated the stresses around holes in infinite sheets of wood. Their data for spruce and oak is shown in Figures 4.4, 4.5 and 4.6.

Note that with the highly anisotropic spruce the maximum stress rises to 6.37σ when the tensile stress σ is applied parallel to the grain and to 2.04σ when σ is applied across the grain. On the other hand, in the latter case there is a region where the compressive stress rises to 5.14σ. The regions of high stress concentration are very localized. Referring to Figure 4.4, the region where the stress exceeds that for an isotropic plate extends radially a distance of 15% of the radius

[19]C. B. Smith. "Effect of Elliptic or Circular Holes on the Stress Distribution in Plates of Wood or Plywood Considered as Orthotropic Materials," *U.S. Dept. of Agric. For. Prod. Mimeo.*, 1510 (1944).

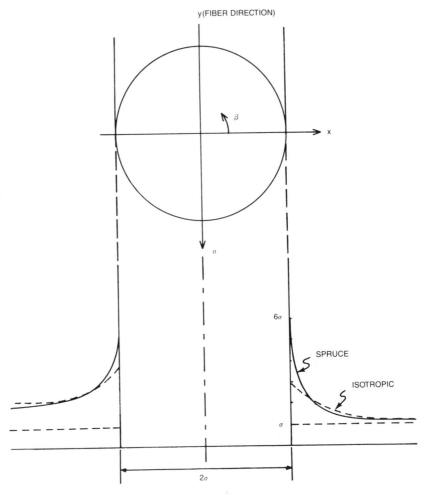

Figure 4.4. Distribution of $\sigma_{\beta\beta}$ along the x-axis ($\beta = 0$). Stress is applied in the fiber direction ($\beta = \pi/2$).

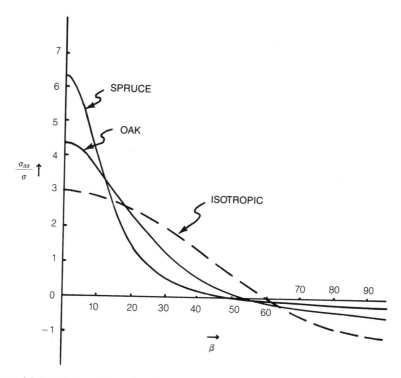

Figure 4.5. Distribution of circumferential stress ($\sigma_{\beta\beta}$ in Inglis' notation) around the edge of a circular hole. Tensile stress σ applied in the fiber direction ($\beta = \pi/2$).

of the hole and, in Figure 4.5, it can be seen that this region extends only 14 degrees from the place of maximum stress concentration. The distribution of shear stress transverse to the fibers is shown in Figure 4.7.

Very similar data have been found for man-made composites, with predicted maximum stress concentrations of $\cong 4$ for glass fiber/epoxy, $\cong 7$ for boron fiber/ epoxy, and $\cong 12$ for high modulus carbon/epoxy uniaxial composites.

In all cases, it is evident that the region of high stress concentration is confined to a small region where the fibers, which have been cut to make the hole, lie close to the uncut fibers. The high stresses which occur near a hole in a stressed plate are of special technological interest because they cause failure of a material which otherwise would have withstood the stress. Since the maximum stress is tensile, failure of the material might be expected to occur at a load smaller by an amount equal to the stress concentration factor, than that which occurs in the absence of the hole. With small holes the material usually withstands greater

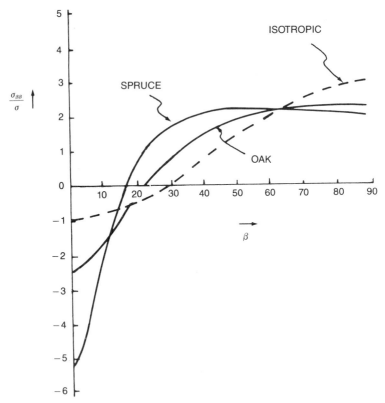

Figure 4.6. Distribution of circumferential stress $\sigma_{\beta\beta}$ (Inglis' notation) around the edge of a circular hole. Tensile stress applied in the transverse direction ($\beta = 0$).

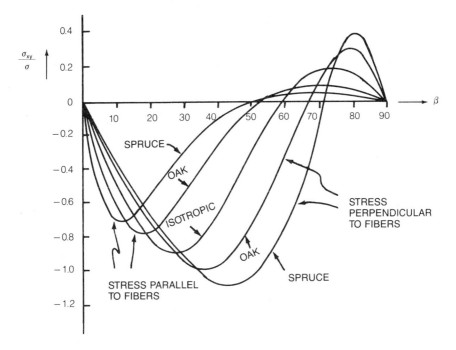

Figure 4.7. Distribution of shear stress σ_{xy} around the edge of a circular hole.

loads, a fact which is usually explained by supposing that the stress at the point of maximum stress concentration does not reach its calculated value because of slight plastic yielding there. This is illustrated by the data plotted in Figure 4.8 for a carbon fiber/epoxy laminate.

In the limit of a hole with diameter equal to one fiber diameter, little if any measurable reduction in strength is expected. The strength corresponding to that predicted by calculation of the stress concentration at the edge of a circular hole is independent of hole size and, in Figure 4.8, is realized only for holes with diameters larger than 5 cm. Between these two limits, the smoothly changing strength is attributed to the nature of the redistribution of load necessitated by cutting of the fibers removed to create the hole.

In general, however, where material has been removed from a laminate in order to provide for windows, inserts and bolt holes, it is good practice to make generous use of doublers.

It is evident from the above observations that the pin through a drilled hole method to grip a tensile test-piece is not an ideal method for fiber reinforced materials. More usual practice is to adhesively bond soft metal (aluminum) tabs to the ends of flat test-pieces and grip these with serrated, self-tightening (tapered) jaws.

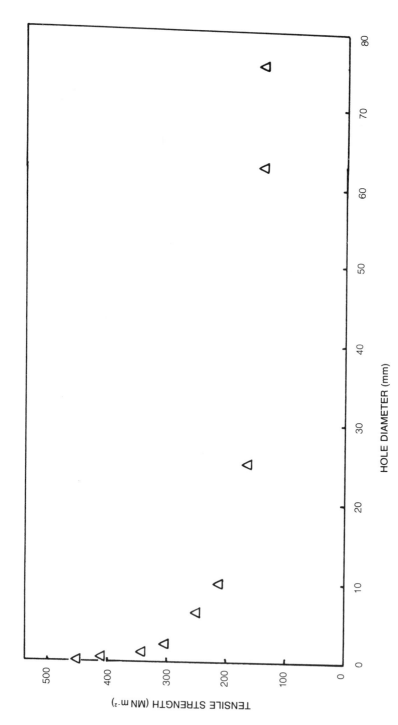

Figure 4.8. Variation of tensile strength with size of circular hole in a $[0/\pm45]_{2S}$ carbon fiber/epoxy resin laminate. [Data taken from Tables 1 and 2 of M. E. Waddoups, J. R. Eisenmann and B. E. Kaminski, "Macroscopic Fracture Mechanics of Advanced Composite Materials," *J. Composite Materials*, 5:446–454 (1971)].

4.3 Free edge effects

The inference from Figure 4.7 that, in a 0° composite tensioned in the fiber direction, σ_{xy}, the magnitude of the in-plane shear stress, can, at the edge of a hole, be a significant fraction of the externally applied stress, has far-reaching consequences; the shear strength of the resin can easily be exceeded here. In the case of laminates, out of plane deformation leading to delamination, that is, to fracture in the resin-rich layer between the plies, can also occur at holes. The origin of stresses at the free edge presented by holes and of stresses at free edges in general has been further investigated by R. B. Pipes and N. J. Pagano.[20] Delamination during fatigue loading is commonly observed at a free edge.

4.4 Worked example[17]

The ultimate strengths (MN m⁻²) for a uniaxial composite, specifically spruce, together with the predominant failure modes in brackets, are as follows.

(a) Axial (longitudinal) tensile strength = 125 (fiber breakages)
(b) Axial (longitudinal) compressive strength = 34.5 (fiber buckling)
(c) Transverse tensile strength = 2.75 to 5.50 (delamination)
(d) Transverse compressive strength = 4.83 (unidentified)
(e) Interlaminar shear strength = 7.60 (interlaminar shear)

Compare the stresses at which failure of the five types might be expected to occur in the neighborhood of a hole during the following deformations.

(i) Axial tension
(ii) Axial compression
(iii) Transverse tension
(iv) Transverse compression

The calculated stress concentration factors referred to the axes shown in Figure 4.9 are listed in Table 4.1. Sigma parallel x, i.e., x,y rotate with sigma.

(i) Axial tension

(a) Maximum tensile value of σ_{xx} occurs at $\theta = 90°$ and is equal to 6.37σ. This corresponds to 125 MN m⁻² so the tensile stress needed to cause failure by fiber breakages is

$$\sigma = \frac{125}{6.37} = 19.6 \text{ MN m}^{-2}$$

(b) Maximum compressive value of σ_{xx} occurs at $\theta = 30°$ and is equal to -0.0308σ. For this to correspond to 34.5 MN m⁻², the applied axial tensile

[20]R. B. Pipes and N. J. Pagano. "Interlaminar Stresses in Composite Laminates Under Uniform Axial Extension," *J. Composite Materials*, 4:538–548 (1970).

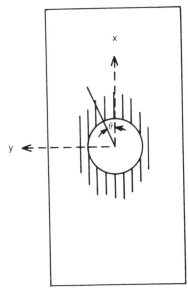

Figure 4.9.

stress would have to be

$$\sigma = \frac{34.5}{0.0308} = 1150 \text{ MN m}^{-2}$$

(c) Maximum tensile value of σ_{yy} occurs at $\theta = 70°$ and is 0.193σ. $\sigma_{yy} = 2.75$ to 5.50 MN m^{-2} would be needed to get delamination, that is

$$\sigma = \frac{2.75 \text{ to } 5.50}{0.193} = 14.25 \text{ to } 28.50 \text{ MN m}^{-2}$$

(d) Maximum compressive value of σ_{yy} occurs at $\theta = 0°$ and is -0.195σ. This corresponds to 4.83 MN m^{-2} so the axial tensile stress necessary to cause failure by this mode is

$$\sigma = \frac{4.83}{0.195} = 24.75 \text{ MN m}^{-2}$$

(e) Maximum value of shear stress σ_{xy} occurs when $\theta = 80°$ and is equal to -0.683σ. Failure by interlaminar shear occurs when $\sigma_{xy} = 7.60$ MN m^{-2}, for which the axial tensile stress would have to be

$$\sigma = \frac{7.60}{0.68} = \boxed{11.20 \text{ MN m}^{-2}}$$

The results, calculated in this way, for the other deformations are:

(ii) Axial compression

(a) $\dfrac{125}{0.0308} = 4058$ MN m^{-2}

(b) $\dfrac{34.5}{6.37} = \boxed{5.40 \text{ MN m}^{-2}}$

(c) $\dfrac{2.75 \text{ to } 5.50}{0.193} = 14$ to 28 MN m^{-2}

(d) $\dfrac{4.83}{0.193} = 25$ MN m^{-2}

(e) $\dfrac{7.60}{0.683} = 11.20$ MN m^{-2}

Table 4.1. Values of stresses on the edge of the circle.

θ	σ_{xy}/σ	σ_{xx}/σ	σ_{yy}/σ
	Tension parallel to grain		
0	0	0	−0.195
10	0.0323	−0.0057	−0.183
20	0.0543	−0.0198	−0.149
30	0.0533	−0.0308	−0.0922
40	0.0137	−0.0115	−0.0164
50	−0.0852	0.102	0.0715
60	−0.266	0.461	0.154
70	−0.529	1.45	0.193
75	−0.653	2.44	0.175
80	−0.683	3.87	0.120
85	−0.483	5.52	0.0423
90	0	6.37	0
	Tension perpendicular to grain		
0	0	0	−5.14
10	0.380	−0.0669	−2.15
20	−0.187	0.0680	0.513
30	−0.735	0.424	1.27
40	−1.02	0.854	1.21
45	−1.06	1.06	1.06
50	−1.06	1.26	0.888
60	−0.922	1.60	0.533
70	−0.671	1.85	0.244
80	−0.352	2.00	0.0621
90	0	2.04	0

(iii) Transverse tension

$$\text{(a)} \quad \frac{125}{1.27} = 98.5 \text{ MN m}^{-2}$$

$$\text{(b)} \quad \frac{34.5}{5.14} = 6.70 \text{ MN m}^{-2}$$

$$\text{(c)} \quad \frac{2.75 \text{ to } 5.50}{2.04} = \boxed{1.35 \text{ to } 2.70 \text{ MN m}^{-2}}$$

$$\text{(d)} \quad \frac{4.83}{0.067} = 71.5 \text{ MN m}^{-2}$$

$$\text{(e)} \quad \frac{7.60}{1.06} = 7.15 \text{ MN m}^{-2}$$

(iv) Transverse compression

$$\text{(a)} \quad \frac{125}{5.14} = 27 \text{ MN m}^{-2}$$

$$\text{(b)} \quad \frac{34.5}{1.27} = 27 \text{ MN m}^{-2}$$

$$\text{(c)} \quad \frac{2.75 \text{ to } 5.50}{0.067} = 41 \text{ to } 82 \text{ MN m}^{-2}$$

$$\text{(d)} \quad \frac{4.83}{2.04} = \boxed{2.35 \text{ MN m}^{-2}}$$

$$\text{(e)} \quad \frac{7.60}{1.06} = 7.15 \text{ MN m}^{-2}$$

The mode of failure which might be expected to occur in each case corresponds with the smallest value for the applied stress. These values have been enclosed with boxes. In cases (ii), (iii) and (iv) failure may be expected at the point on the circumference of the hole which lies on the diameter perpendicular to the direction of the applied stress. This point is also the point of maximum stress when the plate is isotropic. In case (ii) this point corresponds to the point of maximum stress concentration, but in cases (iii) and (iv) there exist greater stresses at other locations. Case (i) is particularly interesting. Here the prediction of failure by interlaminar shear (free-edge effect) is quite different from that predicted for an isotropic plate. The maximum value of σ_{xy} occurs at $\theta = 78°$

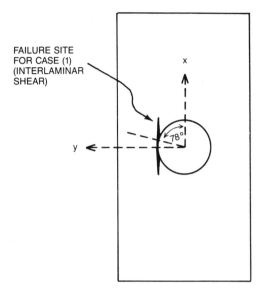

FAILURE SITE
FOR CASE (1)
(INTERLAMINAR
SHEAR)

Figure 4.10.

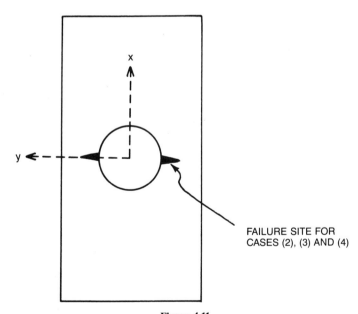

FAILURE SITE FOR
CASES (2), (3) AND (4)

Figure 4.11.

($\beta = 12°$ in Figure 4.7). The calculated location of the shear crack is indicated in Figure 4.10.

The site predicted for failure in cases (ii), (iii) and (iv) is shown in Figure 4.11.

4.5 Further examples

1. It is contrary to common sense that a small hole should weaken a metal just as much as a large one. On the other hand, it is true that a large hole weakens a fiber reinforced composite more than a small one. Explain.
2. A fine scratch made on the surface of plate glass produces such a local weakness to tension that a fracture along the line of the scratch can be brought about by applied forces which produce in the rest of the plate quite insignificant stresses.[18] Discuss.
3. The greatest tensile stress along the rounded corner of a square hole in a plate subjected to a uniaxial tensile stress (σ) occurs at a point between its intersection with the diagonal and its junction with the flat edge and, if the radius of curvature (ϱ) is small compared with the length of the diagonal ($2l$), its approximate value is

$$\frac{\sigma}{2} \sqrt{\frac{l}{\varrho}} \left\{ 1 + \frac{\sqrt{2l + 2\varrho}}{\sqrt{l} - \sqrt{\varrho}} \right\}$$

Estimate this stress magnification for a window in a commercial aircraft.
4. Consider a small crack or notch springing from the side of an elliptic hole. In this case there is a double magnification of stress. The mean stress σ is concentrated to the value $\sigma[1 + (2\sqrt{a/\varrho})]$ in the neighborhood of the crack, and this is magnified to the value $\sigma[1 + (2\sqrt{a/\varrho})][1 + (2\sqrt{a'/\varrho'})]$ at the end of the crack, where a and ϱ refer to the hole and a' and ϱ' refer to the crack. Use this double magnification effect to explain the weakening of a plate which has been punched with holes, and comment on the advisability of removing the rim round the edge of a punched hole.

5

Anisotropy of elasticity

"...the symmetry of wood (and of any 0° laminate)
approximates to that of a rhombic crystal..."[21]

When a stress σ is applied to a solid, we get a strain ϵ and, if Hooke's law is obeyed,

$$\epsilon = s\sigma$$

where s is the compliance.

Alternatively, if a strain ϵ is imposed on the solid, we get a stress σ and, again if Hooke's law is obeyed,

$$\sigma = c\epsilon$$

where c is the stiffness.

More generally, since all of the tensor components of stress or strain are applied

$$\epsilon_{ij} = s_{ijkl}\sigma_{kl} \tag{5.1}$$

and

$$\sigma_{ij} = c_{ijkl}\epsilon_{kl} \tag{5.2}$$

The s_{ijkl} and c_{ijkl} each contain $9 \times 9 = 81$ components.

Physically, what Equation (5.1) tells us is that any one component of strain is not proportional to the corresponding component of stress but is proportional to all of the components of stress, and this proportionality is linear. In fiber lami-

---

[21]H. Horig. "Zür Elastizität des Fichtenholzes. I–Folgerungen aus Messungen von H. Carrington an Spruce," *Zeitschrift für Technische Physik*, 12:369–379 (1931).

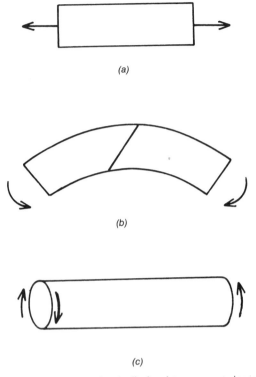

Figure 5.1. (a) Uniaxial stress generates longitudinal and transverse strains, and three shears. (b) Bending produces shears and twists. (c) Twisting produces bending as well as shears.

nates this phenomenon is accentuated by the low symmetry of some lay-ups and is referred to as "coupling" in the fiber reinforced laminates literature.

Experimentally, a uniaxial stress [Figure 5.1(a)] produces strain longitudinally, and transversely in two directions, and three shears. Similarly, in bending [Figure 5.1(b)], application of a couple produces shears and twists. By the same token, in torsion [Figure 5.1(c)], application of a torque generates bending as well as twisting. All of these experimental examples are demonstrations of what Equations (5.1) and (5.2) say.

A noteworthy consequence of torsion-bending coupling in fiber reinforced composites is the effect of longitudinal stiffness on torsional stiffness. In Figure 5.2, consider the effect of fiber orientation on the torsional stiffness of, say, a unidirectionally fiber reinforced compressor fan blade. The shear in plane yz is larger for (b) than for (a). Hence the torsional stiffness about axis z is larger for (a) than for (b).

The s_{ijkl} and c_{ijkl} are fourth rank tensors, the transformation laws for which are of the form

$$s'_{ijkl} = a_{im}a_{jn}a_{ko}a_{lp} \, s_{mnop} \qquad (5.3)$$

where, as in Chapter 3, the prime denotes "new" axes. There are 81 equations (5.3), each with 81 terms. That is $3^4 \times 3^4 = 6561$ coefficients are involved when transforming axes. However, since both stress and strain are symmetrical second rank tensors, they each have six and not nine independent components. So the 81 compliances, stiffnesses and equations for transformation of axes, reduce to 36.

In Chapter 3, second rank tensors were represented as matrices. Here we have fourth rank tensors, for which we would need 4-dimensional matrices. Instead, it is more convenient to change notation from the fourth rank tensor notation to so-called two suffix (or matrix) notation so that compliances and stiffnesses can be represented by 2-dimensional matrices.

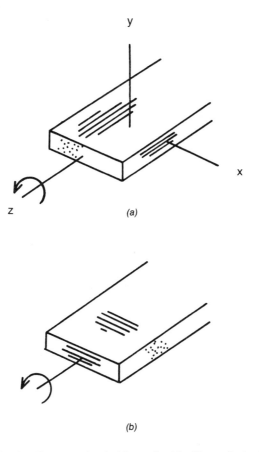

Figure 5.2. Illustrating the effect on torsional stiffness of axial stiffness of a fan blade (after K. T. Kedward).

Write

$$\begin{bmatrix} \sigma_{11} & \sigma_{12} & \sigma_{13} \\ & \sigma_{22} & \sigma_{23} \\ & & \sigma_{33} \end{bmatrix}$$

as

$$\begin{bmatrix} \sigma_1 & \sigma_6 & \sigma_5 \\ & \sigma_2 & \sigma_4 \\ & & \sigma_3 \end{bmatrix}$$

and

$$\begin{bmatrix} \epsilon_{11} & \epsilon_{12} & \epsilon_{13} \\ & \epsilon_{22} & \epsilon_{23} \\ & & \epsilon_{33} \end{bmatrix}$$

as

$$\begin{bmatrix} \epsilon_1 & \tfrac{1}{2}\epsilon_6 & \tfrac{1}{2}\epsilon_5 \\ & \epsilon_2 & \tfrac{1}{2}\epsilon_4 \\ & & \epsilon_3 \end{bmatrix}$$

then

$$s_{ijkl} = s_{mn} \text{ if } m \text{ and } n \text{ are 1, 2 or 3}$$
$$2\,s_{ijkl} = s_{mn} \text{ if either } m \text{ or } n \text{ are 4, 5 or 6} \qquad (5.4)$$
$$4\,s_{ijkl} = s_{mn} \text{ if both } m,\ n \text{ are 4, 5 or 6}$$

Consider some typical equations, e.g., for ϵ_{11} we have

$$\epsilon_1 = \text{set of 9 terms}$$

viz

$$\epsilon_1 = s_{1j}\sigma_j$$

For ϵ_{23} we have

$$\epsilon_4 = \epsilon_{4j}\sigma_j$$

That is

$$\epsilon_i = s_{ij}\sigma_j \qquad (5.5)$$

where $i,j = 1, 2 \ldots 6$

Similarly for the stiffness, but with no factors of 2 and 4 in Equations (5.4)

$$\sigma_i = c_{ij}\epsilon_j \tag{5.6}$$

The upshot is that we have a 6×6 array $= 36$ numbers for the compliance of a material and a different 6×6 array $= 36$ numbers for the stiffness.

However, each array is symmetrical about the leading diagonal. That is

$$c_{ij} = c_{ji}$$

and

$$s_{ij} = s_{ji}$$

Hence the 36 reduces to 21 independent components for each of the compliance and stiffness matrices.

This is the case if the laminate has no particular symmetry. In practice, laminates usually possess symmetry akin to crystal symmetry, and this symmetry further reduces the number of independent constants.

5.1 Elasticity of wood, and of thick 0° laminates in general

H. Horig,[21] in his publication on the elasticity of spruce, pointed out that the symmetry of wood is that of a rhombic (viz. orthorhombic) crystal and that W. Voight's[22] analysis for this crystal system is therefore appropriate. The same is true for a 0° laminate manufactured by laying up parallel strips of prepreg, where the fiber packing through the laminate thickness generally differs from that across the laminate. (Note that a thick uniaxial composite manufactured by resin impregnation of a close packed bundle of fibers has higher symmetry, ideally the symmetry of a hexagonal crystal.)

The characteristic symmetry elements for the orthorhombic crystal system are three mutually perpendicular diad axes of rotation (2 2 2). By way of illustration of the effect of orthorhombic symmetry, consider s_{15} in a uniaxial composite. Converting to tensor notation,

$$s_{15} \text{ becomes } 2\, s_{1131}$$

from which it can be seen that it relates ϵ_{11} and σ_{31} (ϵ_1 and σ_5 in shorthand notation). Suppose we, arbitrarily, set the "through the thickness" direction parallel to the Ox_3 axis. Referring to Figure 5.3, the broken arrows indicate the extension ϵ_{11} generated by application of shear stress σ_{31}. Note also in Figure 5.3 the existence of σ_{13}; the couple due to σ_{31} would tend to rotate the specimen, and the fact that

[22]W. Voight. *Lehrbuch der Kristallphysik*, Druck und Verlag, Leipzig and Berlin, p. 585, pp. 758–760 (1910).

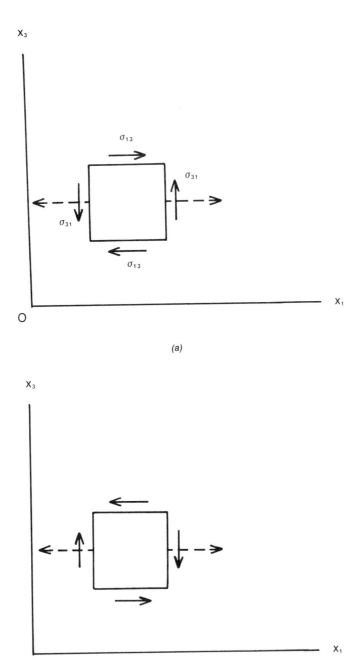

(a)

(b)

Figure 5.3. (a) Illustrating the extension ϵ_{11} (broken arrows) produced by shear stress σ_{31}. Notice the convention that the σ_{31} and σ_{13} furthest from the origin point into the first quadrant for positive shear stress. (b) Reversal of shear stresses by operation of two-fold symmetry.

there occurs strain and not rotation implies the simultaneous presence of σ_{13}. This conclusion is quite general. To answer the question "does the symmetry of the composite allow this?" consider what happens if we rotate the composite, and the stresses applied to it, by π about an Ox_3 axis passing through the center of our unit cube. The composite exhibits two-fold symmetry about Ox_3 so our unit cube looks the same as it did before rotation and so does the extension ϵ_{11}. However, as shown in Figure 5.3(b), the stresses have been reversed. Thus, in the Hooke's law equation $\epsilon = s\sigma$, ϵ still has the same sign but σ has changed sign! Since this is physically impossible, it has to be concluded that s_{15} must be zero.

By considering in this way the effect of symmetry on individual components, or by direct inspection (see J. F. Nye[15]), it is found that nine independent coefficients are needed to completely specify the elastic behavior of an orthorhombic crystal, and hence of a uniaxial composite. The nine coefficients are, for compliance,

$$
s_{ij} = \begin{bmatrix}
s_{11} & s_{12} & s_{13} & 0 & 0 & 0 \\
 & s_{22} & s_{23} & 0 & 0 & 0 \\
 & & s_{33} & 0 & 0 & 0 \\
 & & & s_{44} & 0 & 0 \\
 & & & & s_{55} & 0 \\
 & & & & & s_{66}
\end{bmatrix}
$$

If we, arbitrarily, identify the fiber direction (the direction of the grain in a plank of wood) with the Ox_1 axis, the transverse direction with the Ox_2 axis, and the "through the thickness" direction (the radial direction in the tree trunk from which a plank of wood is sawn) with the Ox_3 axis, then

$$
\left.
\begin{aligned}
s_{11} &= \frac{1}{E_1} = \frac{1}{E_L} \;; & s_{21} &= \frac{-\nu_{21}}{E_2} = \frac{-\nu_{TL}}{E_T} = \frac{-\nu_{12}}{E_1} = \frac{-\nu_{LT}}{E_L} \\[2mm]
s_{31} &= \frac{-\nu_{31}}{E_3} = \frac{-\nu_{RL}}{E_R} = \frac{-\nu_{13}}{E_1} = \frac{-\nu_{LR}}{E_L} \;; & s_{22} &= \frac{1}{E_2} = \frac{1}{E_T} \\[2mm]
s_{32} &= \frac{-\nu_{32}}{E_3} = \frac{-\nu_{RT}}{E_R} = \frac{-\nu_{23}}{E_2} = \frac{\nu_{TR}}{E_T} \;; & s_{33} &= \frac{1}{E_3} = \frac{1}{E_R} \\[2mm]
s_{44} &= \frac{1}{G_{32}} = \frac{1}{G_{RT}} \;; \quad s_{55} = \frac{1}{G_{31}} = \frac{1}{G_{RL}} \;; & s_{66} &= \frac{1}{G_{12}} = \frac{1}{G_{LT}}
\end{aligned}
\right\} \quad (5.7)
$$

where E_L, E_T and E_R are the Young's modulus in the fiber direction, the transverse direction, and the "through the thickness" direction respectively; G_{RT} is the shear modulus for shear stresses in the "through the thickness" and transverse

Table 5.1. Elastic constants of some common woods and of some thick 0° laminates (E and G in [N m^{-2}]).

	$E_L(E_1)$	$E_R(E_3)$	$E_T(E_2)$	$\nu_{RT}(\nu_{32})$	$\nu_{RL}(\nu_{31})$	$\nu_{TR}(\nu_{23})$	$\nu_{TL}(\nu_{21})$	$\nu_{LR}(\nu_{13})$	$\nu_{LT}(\nu_{12})$	$G_{LT}(G_{12})$	$G_{LR}(G_{13})$	$G_{TR}(G_{23})$
Sitka spruce	1.16	9×10^{-1}	5×10^{-2}	0.43	0.029	0.25	0.02	0.37	0.47	7.2×10^{-2}	7.5×10^{-2}	3.9×10^{-3}
Beech	1.37	2.24×10^{-1}	1.14×10^{-1}	0.75	0.073	0.36	0.044	0.45	0.51	1.06×10^{-1}	1.61×10^{-1}	4.6×10^{-2}
Glass fiber/epoxy ($\eta = 3/5$)	43×10^9	—	11.5×10^9	—	—	—	0.072	—	0.27	4×10^9	—	—
Carbon fiber/epoxy ($\eta = 1/2$)	138×10^9	—	14.5×10^9	—	—	—	0.022	—	0.21	6.5×10^9	—	—
Carbon fiber/epoxy ($\eta = 3/5$)	145×10^9	—	10.3×10^9	—	—	—	0.02	—	0.3	6.8×10^9	—	3.5×10^9
Carbon fiber/epoxy ($\eta = 3/5$)	290×10^9	—	6×10^9	—	—	—	0.007	—	0.32	4.5×10^9	—	—
Aramid fiber/epoxy ($\eta = 11/20$)	61×10^9	—	4.1×10^9	—	—	—	0.03	—	0.435	1.54×10^9	—	1.5×10^9
Boron fiber/epoxy ($\eta = 1/2$)	207×10^9	—	14.5×10^9	—	—	—	0.015	—	0.21	5.5×10^9	—	—
Carbon fiber/aluminum ($\eta = 1/2$)	124×10^9	—	25×10^9	—	—	—	0.060	—	0.30	22×10^9	—	—
Carbon fiber/magnesium ($\eta = 2/5$)	90×10^9	—	—	—	—	—	—	—	—	—	—	—

Table 5.2. Elastic compliances of some common woods and of some thick 0° laminates (10^{-12} m² N⁻¹).

	s_{11}	s_{22}	s_{33}	s_{44}	s_{55}	s_{66}	s_{23}	s_{13}	s_{12}
Oregon pine	1,110	61	770	1,110	13,400	1,190	−24	−460	−24.6
Red beech	878	72.6	447	640	2,250	965	−32.7	−325	−37.9
Glassfiber/epoxy ($\eta = 3/5$)	23.5	85.5	—	—	—	242	—	—	−6.5
Carbon fiber/epoxy ($\eta = 1/2$)	7.2	70	—	—	—	153	—	—	−1.5
Boron fiber/epoxy ($\eta = 1/2$)	5	69	—	—	—	180	—	—	−1
Carbon fiber/aluminum	8	40	—	—	—	45	—	—	−2.5

directions, G_{RL} the shear modulus for shear stresses in the "through the thickness" and fiber directions, and G_{LT} the shear modulus for shear stresses in the fiber and transverse directions: ν_{LT} is the major Poisson's ratio, that is

$$\nu_{LT} = \nu_{12} = \frac{\text{contraction in the transverse direction}}{\text{extension in the fiber direction}}$$

for tension applied in the fiber direction. ν_{TR} is the minor Poisson's ratio, that is

$$\nu_{TR} = \nu_{23} = \frac{\text{contraction in the "through the thickness direction"}}{\text{extension in the transverse direction}}$$

for tension applied in the transverse direction.

Table 5.1 lists typical values for some of the elastic constants of a soft wood (spruce), a hard wood (beech), a glassfiber reinforced epoxy resin, three carbon fiber/epoxy composites, an aramid fiber reinforced epoxy resin, a boron fiber reinforced epoxy resin, and two carbon fiber reinforced metals. The respective compliances are given in Table 5.2. The data for the woods are taken from R. F. S. Hearmon.[23] The data for the laminates come from prepreg manufacturers' published design allowables.

The transformation law for fourth rank tensors, Equation (5.3), shows the effect on s_{ij} of rotation from principal axes. For the counterclockwise rotation through angle θ about principal axis Ox_3, shown in Figure 5.4, the following equations give the coefficients of compliance referred to new axes for a material with orthorhombic symmetry.

[23]R. F. S. Hearmon. "The Elasticity of Wood and Plywood," *DSIR Forest Products Research Special Report No. 7*, published by HMSO, London (1948).

$$s'_{11} = s_{11}m^4 + (2s_{21} + s_{66})n^2m^2 + s_{22}n^4 = \frac{1}{E'_1}$$

$$s'_{22} = s_{11}n^4 + (2s_{21} + s_{66})n^2m^2 + s_{22}m^4 = \frac{1}{E'_2}$$

$$s'_{21} + (s_{11} + s_{22})m^2n^2 + s_{21}(m^4 + n^4) - s_{66}m^2n^2 = \frac{\nu'_{21}}{E'_2}$$

$$s'_{66} = 4(s_{11} + s_{22} - 2s_{21})m^2n^2 + s_{66}(m^2 - n^2) = \frac{1}{G'_{12}}$$

$$s'_{61} = -2(s_{11}m^2 - s_{22}n^2)mn + (2s_{21} + s_{66})mn(m^2 - n^2)$$

$$s'_{62} = -2(s_{11}n^2 - s_{22}m^2)mn - (2s_{21} + s_{66})mn(m^2 - n^2)$$

$$s'_{33} = s_{33} = \frac{1}{E_3}$$

$$s'_{44} = s_{44}m^2 + s_{55}n^2 = \frac{1}{G'_{23}}$$

$$s'_{55} = s_{44}n^2 + s_{55}m^2 = \frac{1}{G'_{31}}$$

$$s'_{54} = (s_{44} - s_{55})mn$$

$$s'_{32} = s_{32}m^2 + s_{31}n^2 = \frac{-\nu'_{32}}{E_3}$$

$$s'_{31} = s_{32}n^2 + s_{31}m^2 = \frac{-\nu'_{31}}{E_1}$$

$$s'_{63} = 2(s_{22} - s_{31})mn$$

$$\left. \right\} \quad (5.8)$$

where $m = \cos\theta$, $n = \sin\theta$.

Notice that four new coefficients have appeared, s'_{61}, s'_{62}, s'_{63} and s'_{54}. They are not independent coefficients but can be expressed in terms of the independent coefficients s_{ij} and the angle θ. In accordance with Equation (5.5), s'_{61} relates ϵ'_6, σ'_1, that is ϵ'_{12}, σ'_{11}; s'_{62} relates ϵ'_{12}, σ'_{22}; s'_{63} relates ϵ'_{12}, σ_3; s'_{54} relates ϵ'_{31}, σ_{23}. Thus,

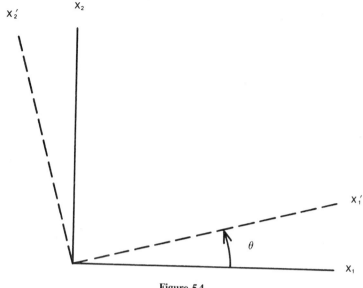

Figure 5.4

normal stresses in the Ox_1', Ox_2', Ox_3 directions generate shears along Ox_1', Ox_2' and conversely; and shear stresses along Ox_2', Ox_3 produce extension in the Ox_3, Ox_1' directions and conversely.

When the elastic constants are inserted into the first of Equations (5.8), we get

$$s_{11}' = \frac{1}{E'} = \frac{1}{E_1} \cos 4\theta + \left(\frac{1}{G_{12}} - \frac{2\nu_{12}}{E_1} \right) \cos 2\theta \sin 2\theta + \frac{1}{E_2} \sin 4\theta$$

the reciprocal of which gives the variation of Young's modulus with angle of rotation θ in the $Ox_1 Ox_2$ plane from Ox_1 towards Ox_2 about Ox_3 as axis. This variation is shown graphically in Figure 5.5(a). The effect of fiber direction on Young's modulus in the $Ox_1 Ox_3$ and $Ox_3 Ox_2$ planes is shown in Figures 5.5(b) and (c).

Figure 5.6 shows the variation with fiber direction of s_{ij} for rotations in the $Ox_1 Ox_2$ plane compiled by R. F. S. Hearmon[23] for both an elastically hard wood (full lines) and an elastically soft wood (broken lines). Two observations are of special interest. Firstly, the change in sign of s_{21}' implies the existence of a negative Poisson's ratio for some fiber directions. Secondly, the relative magnitudes of s_{62}', s_{63}' and s_{54}' which, at angles other than 0° and 90° are as large as the other s_{ij}, show that they cannot be neglected.

The reciprocals of the expressions for s_{55}' and s_{66}' [Equations (5.8)] give the theoretical variation of the shear moduli G_{31}' and G_{12}' with fiber direction. The respective variations are shown graphically in Figures 5.7 and 5.8 for a relatively

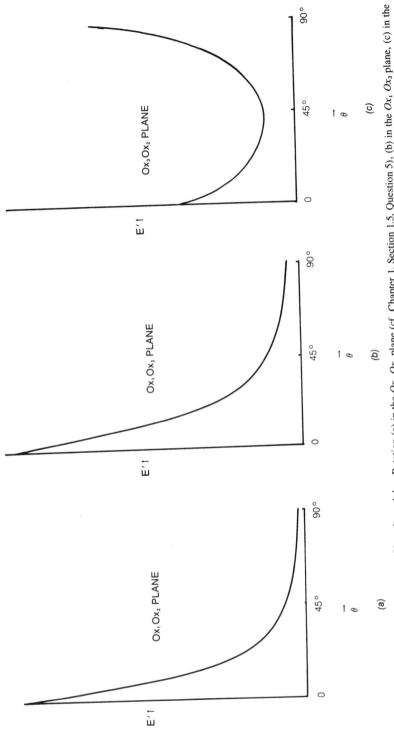

Figure 5.5. Effect of fiber direction on Young's modulus. Rotation (a) in the $Ox_1 Ox_2$ plane (cf. Chapter 1, Section 1.5, Question 5), (b) in the $Ox_1 Ox_3$ plane, (c) in the $Ox_3 Ox_2$ plane (enlarged scale).

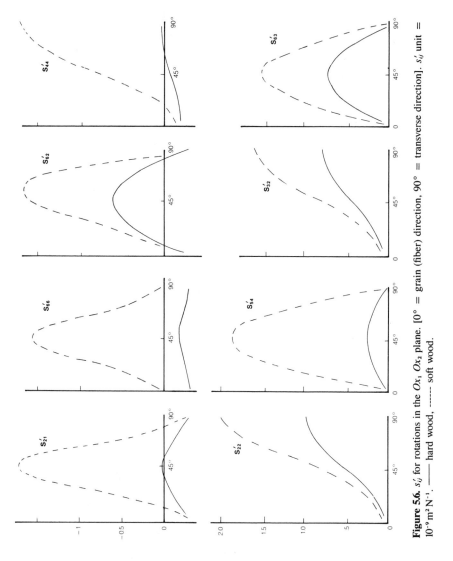

Figure 5.6. s'_{ij} for rotations in the Ox_1, Ox_2 plane. [0° = grain (fiber) direction, 90° = transverse direction]. s'_{ij} unit = 10^{-9} m² N⁻¹. —— hard wood, ----- soft wood.

125

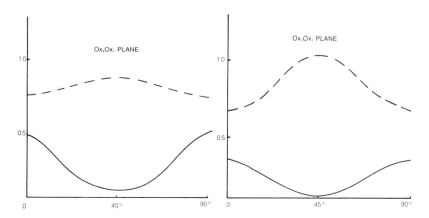

Figure 5.7 Variations of G'_{12} with fiber direction for an elastically strong uniaxial composite (full lines) and an elastically weak uniaxial composite (broken lines). Units = GN m^{-2}. After R. F. S. Hearmon.[23]

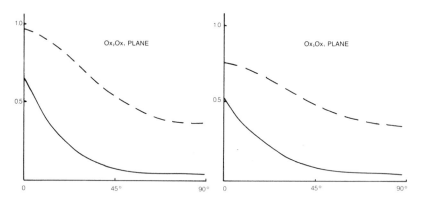

Figure 5.8. Variations of G'_{13} with fiber direction for strong and weak composites. Units = GN m^{-2}. After R. F. S. Hearmon.[23]

hard (elastically) wood (full lines; in units of $10^{-12}\text{m}^2 \text{ N}^{-1}$, $s_{11} = 878$, $s_{22} = 72.6$, $s_{33} = 447$, $s_{44} = 640$, $s_{55} = 2250$, $s_{66} = 965$, $s_{23} = -32.7$, $s_{13} = -325$, $s_{12} = -37.9$) and for an elastically soft wood (broken lines).

5.2 The effect of coupling between shear and extension, and between bending and torsion

Equations (5.8) show that at angles between $0°$ and $90°$ s'_{16} is finite. As pointed out in Section 5.1, this fact implies that tension along Ox'_1 produces a shear strain in the $Ox'_1 Ox'_2$ plane. One physical consequence of this strain is that there is a difference between so-called "pure" and "free" elastic constants. Thus, in the case of Young's modulus, the pure constant is the ratio (extensional stress/extensional strain) measured under conditions for which the induced shear strain is suppressed by simultaneous application of appropriate shear stress. The strain is therefore purely extensional, but a combination of tensile and shear stress is required to produce it. The free Young's modulus is the same ratio measured when the shear strain is allowed to take place. Thus, the stress is purely tensile, but the strain is a combination of extension and shear.

Bending and torsion interact in a similar way. A uniaxial composite for which s'_{16} is non-zero will both twist and bend under the action of either a twisting couple or a bending moment. The pure Young's modulus corresponds to the simultaneous application of a bending moment and a twisting couple such that the deformation is purely flexural. The free Young's modulus corresponds to the application of a simple bending moment so that the deformation is a combination of torsion and flexure.

It can be shown from Equations (5.8) that if the sign of θ is changed, the sign of s'_{11} and s'_{66} remains the same, but the sign of s'_{16} is changed. Physically, this means that if two plies at ply angles θ and $-\theta$ are bonded together the two s'_{16} values countervail each other and the Young's modulus of the $\pm\theta$ laminate is pure whereas that of each individual ply is free. These effects are illustrated for three two-ply strips in Figure 5.9. The fiber direction in the two plies of the upper strip are parallel and are inclined to the specimen axis by $+\theta$ (indicated by the arrow). The fiber direction in the two plies of the lower strip are likewise parallel, but are inclined to the specimen axis by $-\theta$. The strip in the middle is also two plies thick and is subjected to the same end load as the others, but in one of its plies the fiber direction is at angle $+\theta$ and in the other it is at angle $-\theta$ to the specimen axis. Note (i) the twist accompanying bending in the parallel fiber laminates, (ii) the reversal of sign of the twist on reversing the fiber angle, (iii) the absence of twist in the balanced ($\pm\theta$) laminate, (iv) the smaller end deflection of the balanced laminate.

Figures 5.10 and 5.11 summarize the effects of coupling between shear and extension for two-ply ($\pm\theta$) laminates.

Figure 5.12 shows data for $\pm\theta$ symmetric laminates fabricated from high modulus graphite fiber/epoxy resin ($\eta = 3/5$) prepreg. The subscripts x, y, z

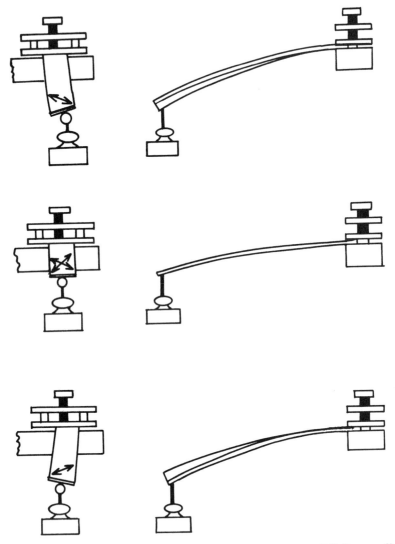

Figure 5.9. Illustrating coupling between bending and twisting. After R. F. S. Hearmon.[23]

128

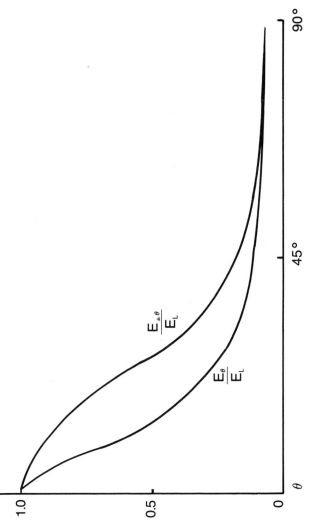

Figure 5.10. Variation of the axial Young's modulus with orientation angle for single ply and symmetric ($\pm \theta$) laminates.

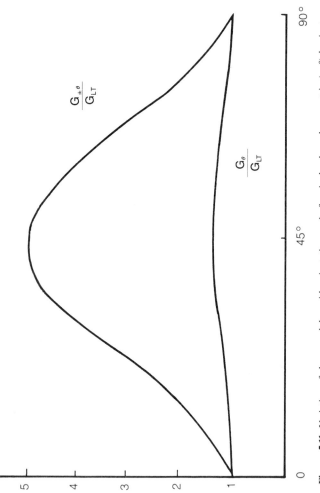

Figure 5.11. Variation of shear modulus with orientation angle for single ply and symmetric ($\pm\theta$) laminates.

130

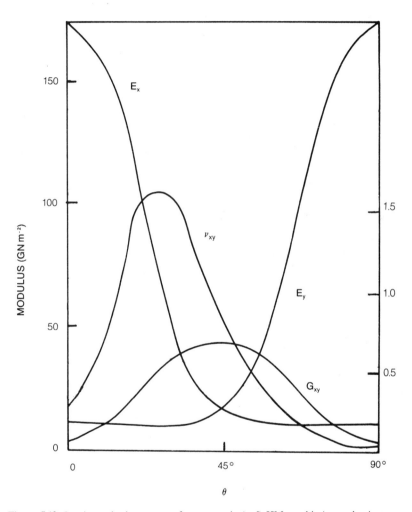

Figure 5.12. Laminate elastic constants for symmetric ($\pm\theta$) HM-graphite/epoxy laminates.

131

Figure 5.13a. Filament wound carbon fiber portable box girder bridge (courtesy of U.S. Army MTL).

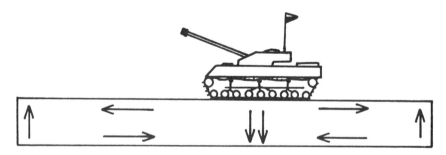

Figure 5.13b. Distribution of stress in (a) when loaded.

refer to the laminate axes (see Chapter 6). Note that $\theta \cong 30°$ for largest shear to extension coupling. Notice also in Figure 5.12 that, as expected from consideration of Figure 3.22, 45° fibers best inhibit shear deformation.

5.3 Worked example

Figure 5.13a is a photograph of a filament wound box girder section for a portable bridge. Explain why the fibers that we see have been wound at ±45° to the length of the bridge. What additional fiber orientation(s) would you recommend for the girders?

In use, the bridge will experience bending and shear. The lower surface of the girder will be in tension, so 0° fibers are required here. 45° fibers will support the shear. Since the sign of the shear in that part of the girder between the load and the north pier will be opposite from that between the load and the south pier, see Figure 5.13b, and since it is intended that the load should move across the bridge, we need ±45° fibers along its whole length.

5.4 Further examples

(1) Verify that the compliance $s_{34} = 0$ in a uniaxial composite.
(2) The coefficients s_{ij} in units of 10^{-12} m^2 N^{-1} for a thick uniaxial composite are as follows: $s_{11} = 4.43$, $s_{22} = 3.53$, $s_{33} = 3.84$, $s_{44} = 9.25$, $s_{55} = 7.52$, $s_{66} = 7.64$, $s_{23} = -0.66$, $s_{13} = -0.86$, $s_{12} = -1.38$. Calculate the compliances referred to axes rotated by 45° about Ox_3.
(3) A three-dimensional composite (cubic symmetry) is subjected to a hydrostatic pressure which produces a dilation of -10^{-6}. What tensile stress parallel to one of the fiber directions must be applied to prevent compression in this direction? $c_{11} = 4.6$ GN m^{-2}, $c_{12} = 3.7$ GN m^{-2}, $c_{44} = 2.6$ GN m^{-2}.
(4) Define (i) the principal axes of strain, and (ii) the principal strains, in a crystal subjected to a small homogeneous deformation.

In a certain orthorhombic crystal $a = 0.31$ nm, $b = 0.42$ nm, $c = 0.67$ nm. If the principal strains are parallel to the crystallographic axes Ox, Oy, Oz, and have the values 3.5×10^{-6}, 7.3×10^{-6}, 0, respectively, what is the extension per unit length of a line in the [122] direction? What is the numerical value of the dilation?

If the deformation is produced by a stress, what can you say about the nature of the stress? What elastic coefficients would be needed to calculate it? Could the deformation be produced in any other way than by a stress (University of Bristol, Examination, 1970)?
(5) How do the elastic constants of a composite depend upon those of its constituents? Give a simple account of two methods of estimation, stating clearly the physical principles involved in each. How accurate would you expect

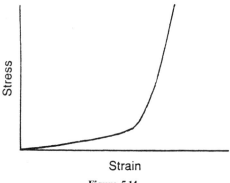

Figure 5.14.

these methods to be in the cases of (a) a unidirectional fiber-reinforced material, and (b) a laminate?

(6) Leather is a natural fiber composite material, made from skins of animals, and is mainly composed of collagen. The stress–strain curve for collagen is sketched above. From your knowledge of the stress–strain behavior of synthetic fibers, suggest reasons why substitutes for leather have never been very satisfactory and indicate ways in which they might be used to better emulate the mechanical behavior of leather.

6

Elasticity of orthotropic laminates

"...the truth is that many professional engineers use very little more than elementary beam theory when they design quite ambitious structures..."
J. E. Gordon (ref. 2, p57)

Thin laminates are used extensively in aerospace construction, as spar webs and wing covering for example. Within the overall structure such components are subjected to loads which usually lie in the laminate plane and produce in-plane compressive stresses and shear stresses.

6.1 Theory of elasticity of a generally-orthotropic plate[24]

Consider a 0/90 laminate. Since it is laid up from plies with fiber directions at right angles to one another, it can be regarded as an *orthogonal anisotropic* plate—known in the notation introduced by M. T. Huber as orthotropic. In the case of identical fiber volume fractions in the two fiber directions, it is evident that a 0/90 laminate has higher symmetry than that of an orthorhombic crystal. The symmetry is, in fact, that of a holosymmetric tetragonal crystal (4/mmm), for which the compliance and stiffness matrices contain only 6 independent components. Equation (5.5) therefore reduces to

$$
\begin{vmatrix} \epsilon_1 \\ \epsilon_2 \\ \epsilon_3 \\ \epsilon_4 \\ \epsilon_5 \\ \epsilon_6 \end{vmatrix}
=
\begin{vmatrix}
\bullet & \bullet & \bullet & 0 & 0 & 0 \\
& \bullet & \bullet & 0 & 0 & 0 \\
& & \bullet & 0 & 0 & 0 \\
& & & \bullet & 0 & 0 \\
& & & & \bullet & 0 \\
& & & & & \bullet
\end{vmatrix}
\begin{vmatrix} \sigma_1 \\ \sigma_2 \\ \sigma_3 \\ \sigma_4 \\ \sigma_5 \\ \sigma_6 \end{vmatrix}
\tag{6.1}
$$

where $\bullet\!\!-\!\!\!-\!\!\bullet$ denotes equal components. If there are no out of plane stresses,

[24]W. Thielmann, "Beitrag zur Frage der Beulung Orthotroper Platten, insbesondere von Sperrholzplatten" (1945), *FGH Report No. 150/19*. For English translation, see U.S. National Advisory Committee for Aeronautics *Technical Memorandum NACA TM 1263* (1950).

135

that is if σ_3 (σ_{33}) $=$ 0, $\sigma_4(\sigma_{23} = \sigma_{32})$ $=$ 0, $\sigma_5(\sigma_{31})$ $=$ 0, then Equation (6.1) becomes

$$
\begin{vmatrix} \epsilon_1 \\ \epsilon_2 \\ \epsilon_3 \\ \epsilon_4 \\ \epsilon_5 \\ \epsilon_6 \end{vmatrix} = \begin{vmatrix} S_{11} & S_{12} & S_{13} & 0 & 0 & 0 \\ & S_{22} & S_{23} & 0 & 0 & 0 \\ & & S_{33} & 0 & 0 & 0 \\ & & & S_{44} & 0 & 0 \\ & & & & S_{55} & 0 \\ & & & & & S_{66} \end{vmatrix} \begin{vmatrix} \sigma_1 \\ \sigma_2 \\ 0 \\ 0 \\ 0 \\ \sigma_6 \end{vmatrix}
$$

where the components within each loop are equal components.
    Hence

$$
\begin{aligned}
\epsilon_1 &= S_{11}\sigma_1 + S_{12}\sigma_2 \\
\epsilon_2 &= S_{21}\sigma_1 + S_{22}\sigma_2 \\
\epsilon_3 &= S_{31}\sigma_1 + S_{32}\sigma_2 \\
\epsilon_4 &= 0 \\
\epsilon_5 &= 0 \\
\epsilon_6 &= S_{66}\sigma_6
\end{aligned}
$$

Since it is a function only of σ_1 and σ_2, ϵ_3 is not independent. The truly independent components of strain are therefore

$$
\begin{vmatrix} \epsilon_1 \\ \epsilon_2 \\ \epsilon_6 \end{vmatrix} = \begin{vmatrix} S_{11} & S_{12} & 0 \\ & S_{22} & 0 \\ & & S_{66} \end{vmatrix} \begin{vmatrix} \sigma_1 \\ \sigma_2 \\ \sigma_6 \end{vmatrix}
$$

In the case considered, a 0/90 laminate with identical fiber volume fractions in the 0 and 90 directions, the symmetry is tetragonal and $s_{11} = s_{22}$. Where this is not the case, and here we include all thin laminates with two-dimensional orthogonal symmetry, $s_{11} \neq s_{22}$. Thus, s_{ij} for a thin orthotropic laminate is an array of four independent components.
    In the same way that the stiffness of a uniaxial composite depends on the orientation of the principal axes of stress in relation to the fiber direction and transverse directions, see Equations (5.8), so too does the stiffness of an orthotropic laminate depend on the orientation of the principal axes of stress in relation to the

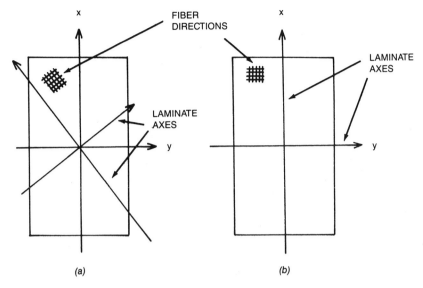

Figure 6.1. (a) Generally-orthotropic plate, (b) specially-orthotropic plate.

main axes of rigidity. This more general case is known as the elasticity of a generally-orthotropic plate for which the s'_{ij}, equations (5.8), are:

$$
\left.
\begin{aligned}
s'_{11} &= s_{11}m^4 + (2s_{12} + s_{66})n^2m^2 + s_{22}n^4 \\
s'_{22} &= s_{11}n^4 + (2s_{12} + s_{66})n^2m^2 + s_{22}m^4 \\
s'_{12} &= (s_{11} + s_{22})m^2n^2 + s_{12}(m^4 + n^4) - s_{66}n^2m^2 \\
s'_{66} &= 4(s_{11} + s_{22} - 2s_{12})m^2n^2 + s_{66}(m^2 - n^2) \\
s'_{16} &= -2(s_{11}m^2 - s_{22}n^2)nm + (2s_{12} + s_{66})nm(m^2 - n^2) \\
s'_{26} &= -2(s_{11}n^2 - s_{22}m^2)nm - (2s_{12} + s_{66})nm(m^2 - n^2)
\end{aligned}
\right\}
\qquad (6.2)
$$

Note that the latter six constants s'_{ij} for the generally-orthotropic plate can be deduced from the four constants s_{ij} for the so-called specially-orthotropic plate, that is there are only four independent constants and these are known as the main directional constants s_{11}, s_{22}, s_{12}, s_{66}.

Generally and specially orthotropic [0/90] laminate plates are illustrated in Figure 6.1(a) and 6.1(b) respectively.

Isotropic materials have only two independent elastic constants because there exists the following relationship between shear modulus (G), the Young's Modulus (E) and Poisson's ratio (ν).

$$
G = \frac{E}{2(1 + \nu)}
\qquad (6.3)
$$

The corresponding relationship for an orthotropic material is

$$s_{66}' = 4(s_{11} + s_{22} - 2s_{12})n^2m^2 + s_{66}(m^2 - n^2) \tag{6.4}$$

which, in the special case of $\theta = 45°$, becomes

$$s_{66}' = s_{11} + s_{22} - 2s_{12} \tag{6.5}$$

What Equation (6.5) says is that when the laminate axes coincide with the principal axes of stress, s_{66}' is independent of s_{66}.

Rewriting Equation (6.5) in terms of G':

$$G' = \cfrac{1}{\left(\cfrac{1}{E_1}\right) + \left(\cfrac{1}{E_2}\right) + \left(\cfrac{\nu_{12}}{E_1}\right) + \left(\cfrac{\nu_{21}}{E_2}\right)}$$

$$= \frac{E_1 E_2}{E_1(1 + \nu_{21}) + E_2(1 + \nu_{12})} \tag{6.6}$$

Similarly, Equations (5.2) become

$$\left.\begin{array}{l} \sigma_x = c_{11}' \epsilon_x + c_{12}' \epsilon_y + c_{16}' \gamma_{xy} \\ \sigma_y = c_{21}' \epsilon_x + c_{22}' \epsilon_y + c_{26}' \gamma_{xy} \\ \tau_{xy} = c_{61}' \epsilon_x + c_{62}' \epsilon_y + c_{66}' \gamma_{xy} \end{array}\right\} \tag{6.7}$$

where the c_{ij}' are

$$\left.\begin{array}{l} c_{11}' = \dfrac{4}{s_{66}} m^2n^2 + \dfrac{s_{11}n^4 + s_{22}m^4 - 2s_{12}m^2n^2}{s_{11}s_{22} - s_{12}^2} \\[3mm] c_{22}' = \dfrac{4}{s_{66}} m^2n^2 + \dfrac{s_{11}m^4 + s_{22}n^2 - 2s_{12}m^2n^2}{s_{11}s_{22} - s_{12}^2} \\[3mm] c_{12}' = - \dfrac{4}{s_{66}} m^2n^2 + \dfrac{(s_{11} + s_{22})m^2n^2 - s_{12}(n^4 + m^4)}{s_{11}s_{22} - s_{12}^2} \\[3mm] c_{66}' = \dfrac{(n^2 - m^2)^2}{s_{66}} + \dfrac{(s_{11} + s_{22} + 2s_{12})m^2n^2}{s_{11}s_{22} - s_{12}^2} \\[3mm] c_{16}' = - mn \dfrac{2}{s_{66}} (n^2 - m^2) + \dfrac{s_{11}n^2 + s_{22}m^2 + s_{12}(n^2 - m^2)}{s_{11}s_{22} - s_{12}^2} \\[3mm] c_{26}' = mn \dfrac{2}{s_{66}} (n^2 - m^2) + \dfrac{s_{11}n^2 + s_{22}m^2 + s_{12}(n^2 - m^2)}{s_{11}s_{22} - s_{12}^2} \end{array}\right\} \tag{6.8}$$

Graphical representations of these equations for glassfiber/epoxy, carbon fiber/epoxy and carbon fiber/aluminum are shown in Figures 6.2, 6.3 and 6.4 respectively. Note the change in notation for stiffness, from c_{ij} used previously, to Q_{ij}, now widely used in the fiber composites literature.

In the case of special-orthotropy, Equations (6.8) are considerably simplified to

$$\left.\begin{aligned}
c_{11} &= Q_{11} = \frac{s_{22}}{s_{11}s_{22} - s_{12}^2} = \frac{E_1}{1 - \nu_{12}\nu_{21}} \\[2em]
c_{22} &= Q_{22} = \frac{s_{11}}{s_{11}s_{22} - s_{12}^2} = \frac{E_2}{1 - \nu_{12}\nu_{21}} \\[2em]
c_{12} &= Q_{12} = \frac{-s_{12}}{s_{11}s_{22} - s_{12}^2} = \frac{\nu_{12}E_2}{1 - \nu_{12}\nu_{21}} = \frac{\nu_{21}E_1}{1 - \nu_{12}\nu_{21}} \\[2em]
c_{66} &= Q_{66} = \frac{1}{s_{66}} = G_{12} \\[2em]
c_{16} &= c_{26} = 0
\end{aligned}\right\} \quad (6.9)$$

Using the elastic constants given in Tables 5.1 and 5.2, the main directional constants Q_{11}, Q_{22}, Q_{12}, Q_{66} for glassfiber/epoxy resin, carbon fiber/epoxy resin and carbon fiber/aluminum are as given in Table 6.1.

Expressed as functions of the main directional stiffnesses, Equations (6.8) on new axes are

$$\left.\begin{aligned}
Q'_{11} &= Q_{11}m^4 + 2(Q_{12} + 2Q_{66})m^2n^2 + Q_{22}n^4 \\
Q'_{22} &= Q_{11}n^4 + 2(Q_{12} + 2Q_{66})m^2n^2 + Q_{22}m^4 \\
Q'_{12} &= (Q_{11} + Q_{22} - 4Q_{66})m^2n^2 + Q_{12}(n^4 + m^4) \\
Q'_{16} &= (Q_{11} - Q_{12} - 2Q_{66})m^3n + (Q_{12} + 2Q_{66} - Q_{22})mn^3 \\
Q'_{26} &= (Q_{11} - Q_{12} - 2Q_{66})mn^3 + (Q_{12} + 2Q_{66} - Q_{22})m^3n \\
Q'_{66} &= (Q_{11} + Q_{22} - 2Q_{12} - 2Q_{66})m^2n^2 + Q_{66}(m^4 + n^4)
\end{aligned}\right\} \quad (6.10)$$

Equations (6.7) for the stresses for specially-orthotropic laminates then run

$$\left.\begin{aligned}
\sigma_x &= \frac{E_1}{1 - \nu_{12}\nu_{21}}\,\epsilon_x + \frac{\nu_{12}E_2}{1 - \nu_{12}\nu_{21}}\,\epsilon_y \\[2em]
\sigma_y &= \frac{\nu_{21}E_1}{1 - \nu_{12}\nu_{21}}\,\epsilon_x + \frac{E_2}{1 - \nu_{12}\nu_{21}}\,\epsilon_y \\[2em]
\tau_{xy} &= G\gamma_{xy}
\end{aligned}\right\} \quad (6.11)$$

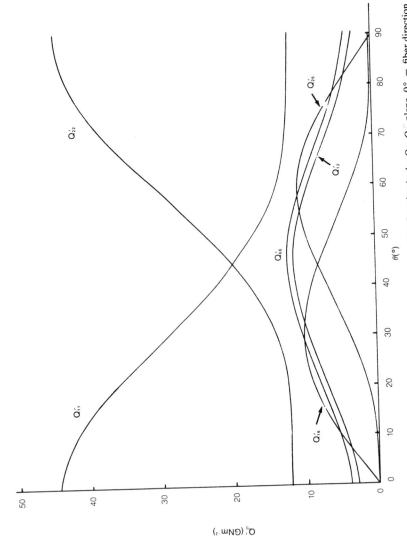

Figure 6.2. $Q'_{ij} (= c'_{ij})$ for glassfiber reinforced epoxy resin ($\eta = 1/2$). Rotation in the $Ox_1 Ox_2$ plane. $0° =$ fiber direction, $90° =$ transverse direction [after S. B. Dong, UCLA (1986)].

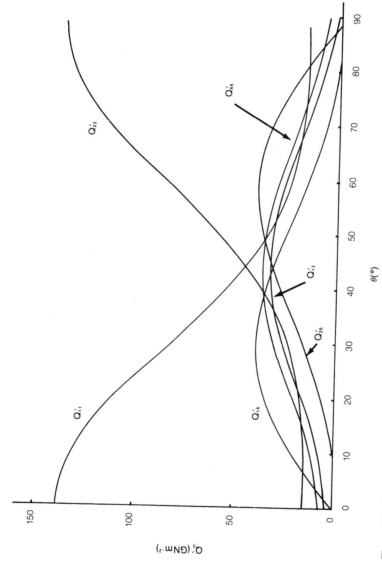

Figure 6.3. Q'_{ij} ($= c'_{ij}$) for carbon fiber reinforced epoxy resin ($\eta = 1/2$). Rotations in the Ox_1, Ox_2 plane. $0°$ = fiber direction, $90°$ = transverse direction [after S. B. Dong, UCLA (1986)].

141

142

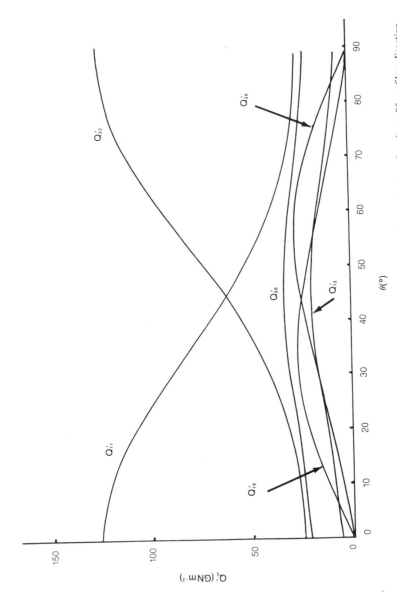

Figure 6.4. $Q'_{ij} (= c'_{ij})$ for carbon fiber reinforced aluminum ($\eta = 2/5$). Rotations in the Ox_1, Ox_2 plane. $0° =$ fiber direction, $90° =$ transverse direction [after S. B. Dong, UCLA (1986)].

Table 6.1. Stiffness coefficients of some thick 0° laminates (GN m^{-2}).

	Q_{11}	Q_{22}	Q_{12}	Q_{66}
glassfiber/epoxy ($\eta = 3/5$)	43.5	12	3	4
carbon fiber/epoxy ($\eta = 1/2$)	140	14.5	3	6.5
boron fiber/epoxy ($\eta = 1/2$)	207	14.5	3	6.5
carbon fiber/aluminum ($\eta = 2/5$)	126	25	7.5	22

In the case of isotropic materials the further simplified constants become

$$c_{11} = c_{22} = \frac{E}{1 - \nu^2}$$

$$c_{12} = \frac{\nu E}{1 - \nu^2} \qquad \left. \right\} \quad (6.12)$$

$$c_{66} = G$$

and the stresses become

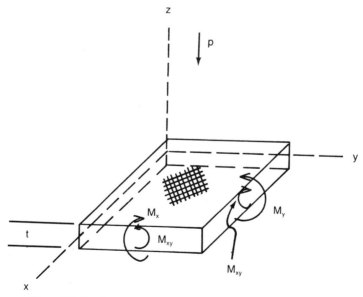

Figure 6.5. Bending moments in an element of orthotropic laminate.

$$\left.\begin{array}{l} \sigma_x = \dfrac{E}{1 - \nu^2}\,(\epsilon_x + \epsilon_y) \\[2em] \sigma_y = \dfrac{E}{1 - \nu^2}\,(\nu\epsilon_x + \epsilon_y) \\[2em] \tau_{xy} = G\gamma_{xy} \end{array}\right\} \qquad (6.13)$$

For the theory of buckling of generally-orthotropic laminates, the usual assumptions of classical theory of bending are made. Above all, it is assumed that the thickness of the laminate is small and that the bending deflection of the laminate itself is small by comparison with the thickness of the laminate.

The bending moments of the laminate, based on the system of coordinate axes shown in Figure 6.5, are

$$\left.\begin{array}{l} M_x = \displaystyle\int_{-t/2}^{t/2} \sigma_x\, z\, dz \\[2.5em] M_y = \displaystyle\int_{-t/2}^{t/2} \sigma_y\, z\, dz \\[2.5em] M_{xy} = \displaystyle\int_{-t/2}^{t/2} \tau_{xy}\, dz \end{array}\right\} \qquad (6.14)$$

The equilibrium condition for the deformation:

$$\frac{\partial^2 M_x}{\partial x^2} + 2\,\frac{\partial^2 M_{xy}}{\partial x \partial y} + \frac{\partial^2 M_y}{\partial y^2} = -p \qquad (6.15)$$

where p is the load acting on the element. This condition is independent of the elastic properties of the plate; it holds for both isotropic and orthotropic plates.

The plate is bent by the moments. Figure 6.6 shows an element in the xz plane. It is assumed that the displacement varies linearly through the plate thickness. To a first approximation, the displacement of point P parallel to Ox is

$$-PP' = -z \tan \alpha$$

$$= -z\,\alpha$$

$$= -z\,\frac{\Delta w}{\Delta x}$$

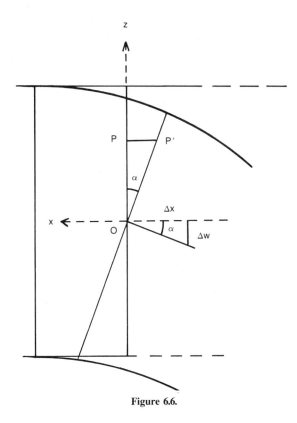

Figure 6.6.

where $\Delta w / \Delta x = \alpha$ is the tilt of the mid-plane and, since strain is the derivative of displacement (see Section 3.11), the corresponding strain is, in the limit,

$$\epsilon_x = -z \frac{\partial^2 w}{\partial x^2}$$

Hence, any point at a distance z from the mid-plane will experience the following components of strain due to the curvature of the plate:

$$\left.\begin{array}{c} \epsilon_x = -z \dfrac{\partial^2 w}{\partial x^2} \\[2ex] \epsilon_y = -z \dfrac{\partial^2 w}{\partial y^2} \\[2ex] \gamma_{xy} = -2z \dfrac{\partial^2 w}{\partial x \partial y} \end{array}\right\} \qquad (6.16)$$

where w denotes the bending deflection of the plate. If the stresses for the generally-orthotropic plate, Equations (6.7), are substituted into Equations (6.14) for the moments of the plate, and if the strains are then replaced by Equations (6.16), integration over the thickness of the plate gives the following expressions for the moments:

$$
\left.
\begin{aligned}
M_x &= \int_{-t/2}^{t/2} (c'_{11}\,\epsilon_x + c'_{12}\,\epsilon_y + c'_{16}\,\gamma_{xy})z\,dz \\
&= -\frac{t^3}{12}\left[c'_{11}\frac{\partial^2 w}{\partial x^2} + c'_{12}\frac{\partial^2 w}{\partial y^2} + 2c'_{16}\frac{\partial^2 w}{\partial x \partial y} \right] \\
M_y &= \int_{-t/2}^{t/2} (c'_{21}\,\epsilon_x + c'_{22}\,\epsilon_y + c'_{26}\,\gamma_{xy})z\,dz \\
&= -\frac{t^3}{12}\left[c'_{21}\frac{\partial^2 w}{\partial x^2} + c'_{22}\frac{\partial^2 w}{\partial y^2} + 2c'_{26}\frac{\partial^2 w}{\partial x \partial y} \right] \\
M_{xy} &= \int_{-t/2}^{t/2} (c'_{61}\,\epsilon_x + c'_{62}\,\epsilon_y + c'_{66}\,\gamma_{xy})z\,dz \\
&= -\frac{t^3}{12}\left[c'_{16}\frac{\partial^2 w}{\partial x^2} + c'_{26}\frac{\partial^2 w}{\partial y^2} + 2c'_{66}\frac{\partial^2 w}{\partial x \partial y} \right]
\end{aligned}
\right\} \quad (6.17)
$$

When the second differentials from Equations (6.17) are substituted into the equilibrium equation, Equation (6.15), we get the differential equation for the elastic surface of the generally-orthotropic plate:

$$
\left.
\begin{aligned}
\frac{t^3}{12}\Bigg\{ &c'_{11}\frac{\partial^4 w}{\partial x^4} + 4c'_{16}\frac{\partial^4 w}{\partial x^3 \partial y} + (2c'_{12} + 4c'_{66})\frac{\partial^4 w}{\partial x^2 \partial y^2} \\
&+ 4c'_{26}\frac{\partial^4 w}{\partial x \partial y^3} + c'_{22}\frac{\partial^4 w}{\partial y^4} \Bigg\} - p = 0
\end{aligned}
\right\} \quad (6.18)
$$

Writing, for abbreviation,

$$ D'_{11} = \frac{t^3}{12} c'_{11} $$

$$ D'_{22} = \frac{t^3}{12} c'_{22} $$

$$D'_{66} = \frac{t^3}{12}(c'_{12} + 2c'_{66})$$

$$D'_{16} = \frac{t^3}{12}c'_{16}$$

$$D'_{26} = \frac{t^3}{12}c'_{26}$$

the differential equation for the buckling load, Equation (6.18), becomes

$$D'_{11}\frac{\partial^4 w}{\partial x^4} + 4D'_{16}\frac{\partial^4 w}{\partial x^3 \partial y} + 2D'_{66}\frac{\partial^4 w}{\partial x^2 \partial y^2} + 4D'_{26}\frac{\partial^4 w}{\partial x \partial y^3} + D'_{22}\frac{\partial^4 w}{\partial y^4} = p$$

(6.19)

In the case of the specially-orthotropic plate,

$$\left.\begin{array}{l} D'_{11} = \dfrac{t^3}{12} \cdot \dfrac{E_{11}}{1 - \nu_{12}\nu_{21}} \\[3mm] D'_{22} = \dfrac{t^3}{12} \cdot \dfrac{E_{22}}{1 - \nu_{12}\nu_{21}} \\[3mm] 2D'_{66} = \dfrac{t^3}{12}\left(\nu_{12}\dfrac{E_{22}}{1 - \nu_{12}\nu_{21}} + \nu_{21}\dfrac{E_{11}}{1 - \nu_{12}\nu_{21}} + 4G\right) \\[3mm] D'_{16} = D'_{26} = 0 \end{array}\right\}$$

(6.20)

and the differential equation reduces to

$$D'_{11}\frac{\partial^4 w}{\partial x^4} + 2D'_{66}\frac{\partial^4 w}{\partial x^2 \partial y^2} + D'_{22}\frac{\partial^4 w}{\partial y^4} = p \qquad (6.21)$$

For an isotropic plate, since

$$E_{11} = E_{22} = E; \; \nu_{12} = \nu_{21} = \nu; \; G = \frac{E(1 - \nu)}{2(1 - \nu^2)}$$

then

$$D'_{11} = D'_{22} = D'_{66} = D' = \frac{t^3}{12} \cdot \frac{E}{(1 - \nu^2)}$$

and the differential equation becomes

$$\frac{\partial^4 w}{\partial x^4} + \frac{2\partial^4 w}{\partial x^2 \partial y^2} + \frac{\partial^4 w}{\partial y^4} = \frac{p}{D} \tag{6.22}$$

which is the well-known $\nabla^4 w$ expression from thin plate elasticity theory for the pressure to deform a membrane supported at its edge (see A. E. H. Love[12] pp. 482–484).

6.2 Constitutive equation for orthotropic laminates

In Figure 6.6 it was assumed that the origin O remains fixed during bending of the laminate. For general deformations, this is not the case; so a vector (u_o, v_o, w_o) representing the displacement of the origin must be added to the displacement equations. Hence Equations (6.16) for the components of strain become

$$\left. \begin{array}{l} \epsilon_x = \dfrac{\partial u_o}{\partial x} - z\,\dfrac{\partial^2 w}{\partial x^2} \\[2ex] \epsilon_y = \dfrac{\partial v_o}{\partial y} - z\,\dfrac{\partial^2 w}{\partial y^2} \\[2ex] \gamma_{xy} = \dfrac{\partial u_o}{\partial y} + \dfrac{\partial v_o}{\partial x} - 2z\,\dfrac{\partial^2 w}{\partial x \partial y} \end{array} \right\} \tag{6.23}$$

that is,

$$\left. \begin{array}{l} \epsilon_x = \epsilon_x^o + z\varkappa_x \\[1ex] \epsilon_y = \epsilon_y^o + z\varkappa_y \\[1ex] \gamma_{xy} = \gamma_{xy}^o + z\varkappa_{xy} \end{array} \right\}$$

where ϵ_x^o, ϵ_y^o are the normal strains, and γ_{xy}^o is the shear strain in the mid-plane, and $\varkappa_x$, $\varkappa_y$, $\varkappa_{xy}$ are the plate curvatures. Equations (6.23) may be rewritten in matrix notation as

$$\begin{bmatrix} \epsilon_x \\ \epsilon_y \\ \gamma_{xy} \end{bmatrix} = \begin{bmatrix} \epsilon_x^o \\ \epsilon_y^o \\ \gamma_{xy}^o \end{bmatrix} + z \begin{bmatrix} \varkappa_x \\ \varkappa_y \\ \varkappa_{xy} \end{bmatrix}$$

or, in shorthand notation,

$$[\epsilon] = [\epsilon^o] + z[\varkappa] \tag{6.24}$$

For an individual ply, the kth ply, in an orthotropic laminate, Hooke's law [Equation (5.2)] becomes

$$[\sigma]_k = [c']_k[\epsilon] \tag{6.25}$$

where the c' are the components of stiffness transformed to the laminate axes. Substituting for $[\epsilon]$ from Equation (6.24) this becomes

$$[\sigma]_k = [c'][\epsilon^0] + z[c'][\varkappa]$$

This, the constitutive equation for a single ply, is usually written in contemporary laminate notation as

$$[\sigma]_k = [\bar{Q}]_k[\epsilon^0] + z[\bar{Q}]_k[\varkappa] \tag{6.26}$$

If the mid-plane strains $[\epsilon^0]$ and the curvatures $[\varkappa]$ are known, Equation (6.26) may be used to calculate the stresses at any point in any ply.
The overall components of stress acting on the laminate are

$$[N] = \sum_{k=1}^{n} \left\{ \int_{t_{k-1}}^{t_k} [\bar{Q}]_k[\epsilon^0]dz + \int_{t_{k-1}}^{t_k} [\bar{Q}]_k [\varkappa]z\, dz \right\} \tag{6.27}$$

and from Equations (6.14), the overall moments are

$$[M] = \sum_{k=1}^{n} \left\{ \int_{t_{k-1}}^{t_k} [\bar{Q}]_k[\epsilon^0]z\, dz + \int_{t_{k-1}}^{t_k} [\bar{Q}]_k [\varkappa]z^2\, dz \right\} \tag{6.28}$$

Since $[\epsilon^0]$ and $[\varkappa]$ are assumed to be uniform through the thickness, Equations (6.27) and (6.28) are reduced to

$$[N] = [A][\epsilon^0] + [B][\varkappa] \tag{6.29}$$

$$[M] = [B][\epsilon^0] + [D][\varkappa] \tag{6.30}$$

where

$$A_{ij} = \sum_{k=1}^{n} (\bar{Q}_{ij})_k(t_k - t_{k-1})$$

$$B_{ij} = \frac{1}{2}\sum_{k=1}^{n} (\bar{Q}_{ij})_k(t_k^2 - t_{k-1}^2) \tag{6.31}$$

$$D_{ij} = \frac{1}{3}\sum_{k=1}^{n} (\bar{Q}_{ij})_k(t_k^3 - t_{k-1}^3)$$

Combining Equations (6.29) and (6.30) yields the laminate constitutive equation

$$\begin{bmatrix} N \\ M \end{bmatrix} = \begin{bmatrix} A & B \\ B & D \end{bmatrix} \begin{bmatrix} \epsilon^o \\ \varkappa \end{bmatrix} \tag{6.32}$$

or, in longhand,

$$\begin{bmatrix} N_x \\ N_y \\ N_{xy} \\ M_x \\ M_y \\ M_{xy} \end{bmatrix} = \left[\begin{array}{ccc|ccc} A_{11} & A_{12} & A_{16} & B_{11} & B_{12} & B_{16} \\ & A_{22} & A_{26} & & B_{22} & B_{26} \\ & & A_{66} & & & B_{66} \\ \hline & & & D_{11} & D_{12} & D_{16} \\ & & & & D_{22} & D_{26} \\ & & & & & D_{66} \end{array} \right] \begin{bmatrix} \epsilon_x^o \\ \epsilon_y^o \\ \gamma_{xy}^o \\ \varkappa_x \\ \varkappa_y \\ \varkappa_{xy} \end{bmatrix} \tag{6.33}$$

Sub-matrices A_{ij}, B_{ij} and D_{ij} are all different summations of the ply stiffnesses, see Equations (6.31). A_{ij} is used to evaluate in-plane stresses. B_{ij} is used to evaluate both in-plane stresses and bending moments. D_{ij} is used to evaluate bending moments.

There is one word of warning, however, when using the sub-matrices in practice, which was pointed out by Thielmann.[24] Ignoring microstructural defects like fiber cross-overs, a single ply can be regarded as a homogeneous orthotropic material to which the theory of anisotropic elasticity and, in particular, the theory of bending can be applied. However, when a number of plies are adhesively bonded together to form a laminate, we may obtain a material which is likewise characterized by orthotropic symmetry but which is certainly not homogeneous through the thickness of the laminate. Even without fabrication defects, it is evident that, if we formally retain the relationships which are valid for a homogeneous material, the elastic moduli obtained for a laminate when stressed in the plane of the laminate will in part assume different values from those obtained when it is bent out of the laminate plane. Consider the $[\theta_1/\theta_2]_s$ laminate shown in Figure 6.7. When stressed uniaxially in the plane of the laminate, e.g., by tension along Ox

$$E_x \cdot e \cdot a \cdot 2t_2 = E_{\theta_1} e \cdot a \cdot 2t_1 + E_{\theta_2} \cdot e \cdot a \cdot 2(t_2 - t_1)$$

where a is the width of the laminate. The strain e is assumed to be uniform over the cross-section. Hence

$$E_x t_2 = E_x 2t_1 = E_{\theta_1} t_1 + E_{\theta_2} (t_2 - t_1) \tag{6.34}$$

and the Young's modulus

$$E_x = \frac{E_{\theta_1} + E_{\theta_2}}{2} \tag{6.35}$$

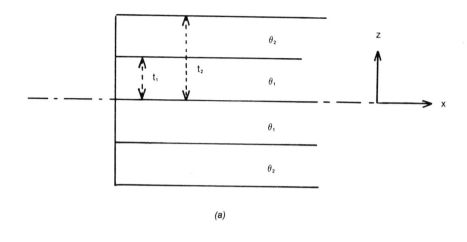

(a)

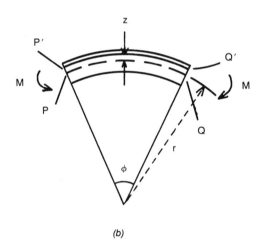

(b)

Figure 6.7.

Now consider bending about axis Oy, Figure 6.7(b), and assume that the strain varies linearly through the thickness of the laminate, that is from its maximum positive value at the tensioned free surface, through zero at the symmetry plane, to its maximum negative value at the compressed free surface.

Suppose the laminate is bent into cylindrical curvature and that the radius of the neutral surface PQ is r. If we consider a filamentary surface $P'Q'$, a distance z from PQ, we have

$$P'Q' = (r + z)\phi$$

Hence the extension of the filamentary surface is

$$e = P'Q' - PQ$$

$$= (r + z)\phi - r\phi$$

$$= z\phi$$

and, since the original length was $r\phi$, the strain is z/r and the force acting across the cross-section is

$$\frac{E_x \cdot z}{r} 2t_2 a$$

The couple due to these forces is thus

$$\frac{E_x \cdot z}{r} 2t_2 a \cdot z$$

and the total couple, or bending moment, due to all the filamentary planes in the plate, which must equal the external applied moment M, will be

$$M = \frac{E_x}{r} \sum 2t_2 \, a \, z^2$$

or, in integral form,

$$M = \frac{E_x}{r} \cdot 2a \int_0^{t_2} z^2 \, dz$$

Rewriting this for our laminate, we have

$$M = \frac{E_{\theta_1}}{r} \cdot 2a \int_0^{t_1} z^2 \, dz + \frac{E_{\theta_2}}{r} \cdot 2a \int_0^{t_2} z^2 \, dz$$

Hence, evaluating the integrals,

$$E_x \, t_2^3 = E_x \, 8t_1^3 = E_{\theta_1} \, t_1^3 + E_{\theta_2} \, (t_2^3 - t_1^3) \tag{6.36}$$

and the Young's modulus

$$E_x = \frac{E_{\theta_1} + 7E_{\theta_2}}{8} \tag{6.37}$$

It is evident from Equations (6.35) and (6.37) that the Young's modulus for the laminate deformed in-plane must, in general, differ from the same constant measured in bending.

Extending Equations (6.34) and (6.36) to the general case of a symmetrical $(2n - 1)$ laminate, gives, for in-plane uniaxial deformation along laminate axis Ox,

$$E_x t_n = E_{\theta_1} t_1 + E_{\theta_2} (t_2 - t_1) + \ldots + E_{\theta_n}[t_n - (t_{n-1})] \qquad (6.38)$$

and, for out of plane deformation by bending about laminate axis Oz,

$$E_x t_n^3 = E_{\theta_1} t_1^3 + E_{\theta_2} (t_2^3 - t_1^3) + \ldots + E_{\theta_n}[t_n^3 - (t_{n-1}^3)] \qquad (6.39)$$

The apparent shear modulus in bending is given by an equation analogous to (6.38) with Young's moduli E replaced by the shear moduli G. The equation for apparent shear modulus in torsion is similarly analogous to Equation (6.39) (cf. Section 5.2 and Figure 5.9).

Figure 6.8, after K. T. Kedward (Texas A&M University, 1985), shows geometrical illustrations of some of the coupling phenomena that can be deduced by inspection of the laminate constitutive Equation (6.33). A_{16}, for example, relates σ_1 (i.e., σ_x), ϵ_6 (i.e., $2 \epsilon_{xy}$); B_{26} relates σ_2 (i.e., σ_y), ϵ_6 (i.e., $2 \epsilon_{xy}$) and also M_2 (i.e., M_y), $\varkappa_6$ (i.e., $\varkappa_{xy}$); D_{66} relates M_6 (i.e., M_{xy}), $\varkappa_6$ (i.e., M_{xy}).

The change in shape illustrated for B_{11} coupling, for example, can be visualized by considering the effect of a tensile stress on an unsymmetric laminate. Thus, a $0°$, $90°$ two-ply laminate tensioned in the $0°$ direction presents to the tensile stress a stiff layer and a soft layer, so that it bends. Similarly, the B_{16} coupling can be visualized by considering an unsymmetric $\pm \theta$ two-ply laminate. Suppose the upper ply has its fiber direction at angle $+\theta$ and that the lower ply has its fiber direction at $-\theta$ to the tensile axis. The two plies shear in the opposite sense, so the laminate twists. D_{16} coupling in symmetric laminates is readily visualized by considering the effect of a bending moment on a $[\pm \theta, \bar{0}]_s$ laminate. Shears parallel to the axis of bending are generated in the angle plies. These shears are of opposite sense and different magnitudes in the two angle ply orientations. Hence the laminate twists.

The consequences of coupling are spectacular in $\pm \theta$ two-ply laminates. However, by the time the number of plies reaches six, the deformation is well approximated by the equations for an orthotropic plate. The same conclusion holds for fundamental bending frequency.

6.3 Carpet plots

One guide for designers that is easily generated by computer codes for laminates is the so-called carpet plot. Carpet plots are graphs which show the variations in elastic moduli of laminates predicted by varying the proportions of the

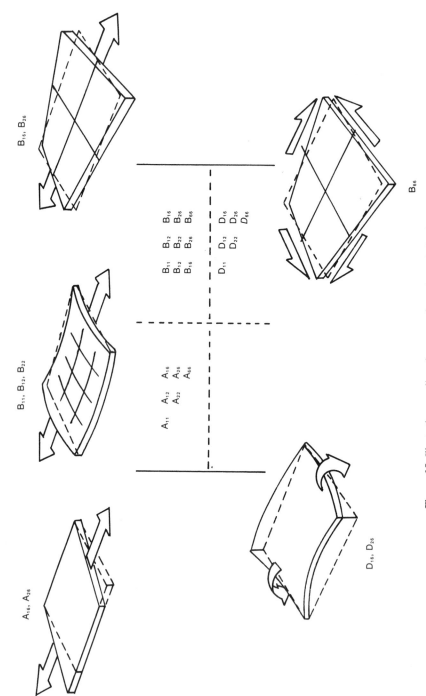

$$
\begin{array}{ccc}
A_{11} & A_{12} & A_{16} \\
 & A_{22} & A_{26} \\
 & & A_{66}
\end{array}
\qquad
\begin{array}{ccc}
B_{11} & B_{12} & B_{16} \\
B_{12} & B_{22} & B_{26} \\
B_{16} & B_{26} & B_{66}
\end{array}
\qquad
\begin{array}{ccc}
D_{11} & D_{12} & D_{16} \\
 & D_{22} & D_{26} \\
 & & D_{66}
\end{array}
$$

A_{16}, A_{26}

B_{11}, B_{12}, B_{22}

B_{16}, B_{26}

D_{16}, D_{26}

B_{66}

Figure 6.8. Illustrating coupling phenomena for orthotropic laminates.

% 0° PLIES

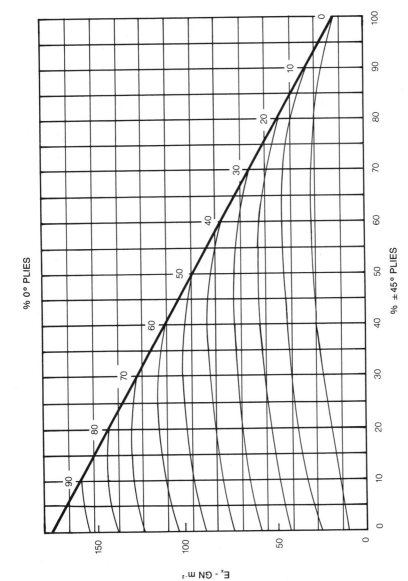

Figure 6.9. Longitudinal modulus (E_x) for $[0_i/\pm45_j/90_k]_S$ carbon fiber/epoxy resin laminates.

% ±45° PLIES

E_x - GN m^{-2}

different ply orientations. Figures 6.9 and 6.10 are carpet plots for longitudinal modulus E_x and major Poisson's ratio ν_{xy} computed for $[0_i/\pm 45_j/90_k]_s$ high modulus carbon fiber/epoxy resin laminates. The subscripts i, j and k denote the numbers of plies with, respectively, $0°$, $\pm 45°$ and $90°$ orientations. Both figures reveal examples of the physical insight that can be gleaned from computer experiments. Details of examples of those are described below.

In Figure 6.9 E_x falls from about 175 GN m^{-2} for the $0°$ laminate ($j=k=0$) to $\cong 20$ GN m^{-2} for the $\pm 45°$ laminate ($i=k=0$). The near horizontal tie-bars show how E_x changes when $\pm 45°$ plies are replaced by $90°$ plies. The left hand end of each tie-bar represents replacements of all ± 45s by 90s, the right hand end the replacement of no ± 45s by 90s. Each tie-bar rises to a maximum between its ends, indicating that the introduction of $\pm 45°$ fibers causes the longitudinal stiffness to increase. The reason for this increase is indirect; it is because the major Poisson's ratio expansion is constrained by the $\pm 45°$ fibers. Referring to Figure 6.10, the upper more-or-less horizontal line shows the increase from $\nu_{xy} = 0.3$ for the $0°$ laminate ($j=k=0$) to $\nu_{xy} \geq 0.8$ for the $\pm 45°$ laminate ($i=k=0$). $\nu_{xy} = 0.8$ is remarkably high; it means that 80% of the deformation is in the transverse direction. Thus, when bent about axis Oy, the curvature about axis Ox exceeds that about Oy. The tie-bars directed downward from the near horizontal line show the effect on ν_{xy} of replacing $\pm 45°$ plies with $90°$plies. The tie-bar slopes are steep indicating that very few $90°$ plies are required to lower ν_{xy} to below 0.5.

6.4 Worked examples

1. Strain gage measurements on a deformed laminate show that $\epsilon_x = 3.5 \times 10^{-3}$, $\epsilon_y = 5 \times 10^{-4}$, $\gamma_{xy} = 5.2 \times 10^{-3}$. The prepreg manufacturer quotes the following as typical data for unidirectional laminate manufactured from his material: $E_{11} = 138$ GN m^{-2}, $E_{22} = 13.8$ GN m^{-2}, $G_{12} = 6.9$ GN m^{-2}, $\nu_{12} = 0.2$. What state of stress is experienced by a $30°$ ply? What is this state of stress referred to the laminate axes (after K. T. Kedward, UCLA, 1980)?

Using the transformation law for second rank tensors, Equation (3.1), we have

$$\epsilon'_{ij} = a_{il}\, a_{jm}\, \epsilon_{lm}$$

For rotation θ about principal axis Ox_3, we have

$$
\begin{bmatrix} \epsilon_1 \\ \epsilon_2 \\ \dfrac{\gamma_{12}}{2} \end{bmatrix}
=
\begin{bmatrix} c^2 & s^2 & 2sc \\ s^2 & c^2 & -2sc \\ -sc & sc & C2 \end{bmatrix}
\begin{bmatrix} \epsilon_x \\ \epsilon_y \\ \dfrac{\gamma_{xy}}{2} \end{bmatrix}
$$

% 0° PLIES

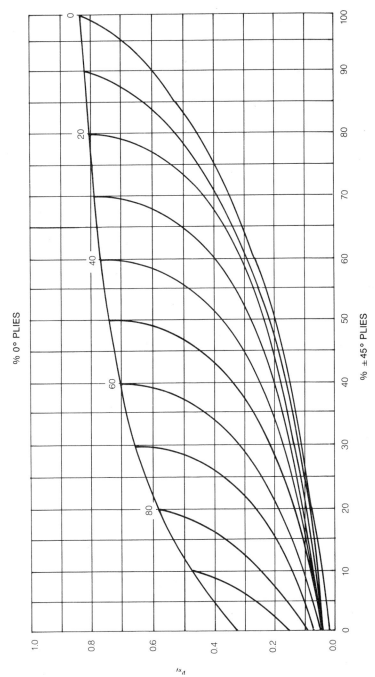

% ±45° PLIES

Figure 6.10. Major Poisson's ratio for $[0_i/\pm45_j/90_k]_S$ carbon fiber/epoxy resin laminates.

where c denotes $\cos\theta$, s denotes $\sin\theta$ and $C2$ denotes $\cos 2\theta$.

$$
\begin{bmatrix} \epsilon_1 \\ \epsilon_2 \\ \dfrac{\gamma_{12}}{2} \end{bmatrix} = \begin{bmatrix} 0.75 & 0.25 & 0.866 \\ 0.25 & 0.75 & -0.866 \\ -0.433 & 0.433 & 0.5 \end{bmatrix} \begin{bmatrix} 3.5 \times 10^{-3} \\ 5.0 \times 10^{-4} \\ \dfrac{5.2}{2} \times 10^{-3} \end{bmatrix}
$$

$$
= \begin{bmatrix} 0.005 \\ -0.001 \\ 0 \end{bmatrix}
$$

Hooke's law, Equation (5.2), for a specially orthotropic laminate is

$$
\begin{bmatrix} \sigma_1 \\ \sigma_2 \\ \tau_{12} \end{bmatrix} = \begin{bmatrix} Q_{11} & Q_{12} & 0 \\ Q_{12} & Q_{22} & 0 \\ 0 & 0 & Q_{66} \end{bmatrix} \begin{bmatrix} \epsilon_1 \\ \epsilon_2 \\ \gamma_{12} \end{bmatrix}
$$

where the components of stiffness are given by Equations (6.9)

$$
Q_{11} = c_{11} = \frac{E_{11}}{1 - \nu_{12}\nu_{21}} = 138 \text{ GN m}^{-2}
$$

$$
\left(\nu_{21} = \frac{E_{22}}{E_{11}} \nu_{12} = 0.02 \right)
$$

$$
Q_{22} = c_{22} = \frac{E_{22}}{1 - \nu_{12}\nu_{21}} = 13.8 \text{ GN m}^{-2}
$$

$$
Q_{12} = c_{12} = \frac{\nu_{21}E_{11}}{1 - \nu_{12}\nu_{21}} = 2.76 \text{ GN m}^{-2}
$$

$Q_{66} = c_{66}$ ($= c_{33}$ in the notation used by Thielmann) $= G_{12} = 6.9$ GN m^{-2}. Hence,

$$
\begin{bmatrix} \sigma_1 \\ \sigma_2 \\ \tau_{12} \end{bmatrix} = \begin{bmatrix} 138 & 2.76 & 0 \\ 2.76 & 13.8 & 0 \\ 0 & 0 & 6.9 \end{bmatrix} \begin{bmatrix} 0.005 \\ -0.001 \\ 0 \end{bmatrix} \text{ GN m}^{-2}
$$

$$= \begin{bmatrix} 687 \\ 0 \\ 0 \end{bmatrix} \text{MN m}^{-2}$$

that is, a 30° ply experiences uniaxial tensile stress of 687 MN m^{-2} in the fiber direction.

Using the reverse transformation law for second rank tensors, Equation (3.2), we have

$$\sigma_{ij} = a_{ki} \, a_{lj} \, \sigma'_{kl}$$

That is

$$\begin{bmatrix} \sigma_x \\ \sigma_y \\ \tau_{xy} \end{bmatrix} = \begin{bmatrix} 0.75 & 0.25 & -0.866 \\ 0.25 & 0.75 & 0.866 \\ 0.433 & -0.433 & 0.5 \end{bmatrix} \begin{bmatrix} 687 \\ 0 \\ 0 \end{bmatrix} \text{MN m}^{-2}$$

$$= \begin{bmatrix} 515 \\ 173 \\ 297 \end{bmatrix} \text{MN m}^{-2}$$

2. The computed coefficients $[A_{ij}]$, see Equations (6.31), for a one inch (25.4 mm) thick symmetric ±45° laminate are as follows:

$$[A_{ij}] = \begin{bmatrix} 71.0 & 52.5 & 0 \\ 52.5 & 71.0 & 0 \\ 0 & 0 & 58.5 \end{bmatrix} \text{GN m}^{-2}$$

The stiffness matrix for a single ply is

$$[Q_{ij}] = \begin{bmatrix} 220 & 3 & 0 \\ 3 & 20 & 0 \\ 0 & 0 & 10 \end{bmatrix} \text{GN m}^{-2}$$

Calculate the ply stresses (a) referred to the ply axes, and (b) referred to the laminate axes when the laminate is subjected to a uniaxial tensile stress $\sigma_x = 150$ MN m^{-2} (after K. T. Kedward, UCLA 1980).

(a) For symmetric laminates there exists no coupling between bending and stretching (see Section 5.2). Hence Hooke's law, Equation (5.2) reduces to

$$\begin{bmatrix} \sigma_x \\ \sigma_y \\ \tau_{xy} \end{bmatrix} = \begin{bmatrix} 150 \\ 0 \\ 0 \end{bmatrix} \times 10^6 \text{ N m}^{-2} = \begin{bmatrix} 71.0 & 52.5 & 0 \\ 52.5 & 71.0 & 0 \\ 0 & 0 & 58.5 \end{bmatrix} \begin{bmatrix} \epsilon_x \\ \epsilon_y \\ \gamma_{yx} \end{bmatrix} \times 10^9 \text{ N m}^{-2}$$

$$150 \times 10^{-3} = 71.0 \, \epsilon_x + 52.5 \, \epsilon_y \quad \text{(i)}$$
$$0 = 52.5 \, \epsilon_x + 71.0 \, \epsilon_y \quad \text{(ii)}$$
$$0 = 58.5 \, \gamma_{xy}$$

from (i) and (ii),

$$(150 \times 10^{-3}) \times 52.5 = (71.0 \times 52.5)\epsilon_x + (52.5)^2\epsilon_y \qquad \text{(iii)}$$
$$0 = (71.0 \times 52.5)\epsilon_x + (71.0)^2\epsilon_y \qquad \text{(iv)}$$

Subtracting (iv) from (iii),

$$7.90 = -2285 \ \epsilon_y$$

so

$$\epsilon_y = -3.46 \times 10^{-3}$$

from (ii),

$$\epsilon_x = 4.68 \times 10^{-3}$$
$$\gamma_{xy} = 0$$

Using the transformation law for second rank tensors, Equation (3.1) to find the strains in the $+45$ ply,

$$\begin{bmatrix} \epsilon_1 \\ \epsilon_2 \\ \dfrac{\gamma_{12}}{2} \end{bmatrix} = \begin{bmatrix} c^2 & s^2 & 2sc \\ s^2 & c^2 & -2sc \\ -sc & sc & C2 \end{bmatrix} \begin{bmatrix} \epsilon_x \\ \epsilon_y \\ \dfrac{\gamma_{xy}}{2} \end{bmatrix}$$

$$= \begin{bmatrix} \frac{1}{2} & \frac{1}{2} & 1 \\ \frac{1}{2} & \frac{1}{2} & -1 \\ -\frac{1}{2} & \frac{1}{2} & 0 \end{bmatrix} \begin{bmatrix} 4.68 \\ -3.46 \\ 0 \end{bmatrix} \times 10^{-3}$$

$$\epsilon_1 = \{(\tfrac{1}{2} \times 4.68) - (\tfrac{1}{2} \times 3.46)\} \times 10^{-3}$$

$$= 0.61 \times 10^{-3}$$

$$\epsilon_2 = \{(\tfrac{1}{2} \times 4.68) - (\tfrac{1}{2} \times 3.46)\} \times 10^{-3}$$

$$= 0.61 \times 10^{-3}$$

$$\frac{\gamma_{12}}{2} = -\{(\frac{1}{2} \times 4.68) - (\frac{1}{2} \times 3.46)\} \times 10^{-3}$$

$$= -4.07 \times 10^{-3}$$

$$\gamma_{12} = -8.14 \times 10^{-3}$$

Using Hooke's law on a single ply

$$\begin{bmatrix} \sigma_1 \\ \sigma_2 \\ \tau_{12} \end{bmatrix} = \begin{bmatrix} 220 & 3 & 0 \\ 3 & 20 & 0 \\ 0 & 0 & 10 \end{bmatrix} \begin{bmatrix} 0.61 \\ 0.61 \\ -8.14 \end{bmatrix} \times 10^{-3} \text{ GN m}^{-2}$$

$$\sigma_1 = \{(220 \times 0.61) + (3 \times 0.61)\} \times 10^{-3} \text{ GN m}^{-2}$$

$$= 135.4 \text{ MN m}^{-2}$$

$$\sigma_2 = \{(3 \times 0.61) + (20 \times 0.61)\} \times 10^{-3} \text{ GN m}^{-2}$$

$$= 14.0 \text{ MN m}^{-2}$$

$$\tau_{12} = -81.4 \text{ MN m}^{-2}$$

These are the stresses in the $+45°$ ply. In the $-45°$ ply, the stresses will be

$$\sigma_1 = 135.4 \text{ MN m}^{-2}$$
$$\sigma_2 = 14.0 \text{ MN m}^{-2}$$
$$\tau_{12} = +81.4 \text{ MN m}^{-2}$$

(b) To transform these stresses to laminate axes, use second rank tensor transformation law. For the $+45°$ ply, we have

$$\begin{bmatrix} \sigma_x \\ \sigma_y \\ \tau_{xy} \end{bmatrix} = \begin{bmatrix} \frac{1}{2} & \frac{1}{2} & -1 \\ \frac{1}{2} & \frac{1}{2} & 1 \\ \frac{1}{2} & -\frac{1}{2} & 0 \end{bmatrix} \begin{bmatrix} 135.4 \\ 14.0 \\ -81.4 \end{bmatrix} \text{ MN m}^{-2}$$

$$\sigma_x = \{(\frac{1}{2} \times 135.4) + (\frac{1}{2} \times 14.0) + 81.4\} \text{ MN m}^{-2}$$

$$= 156.4 \text{ MN m}^{-2}$$

$$\sigma_y = \{(\tfrac{1}{2} \times 135.4) + (\tfrac{1}{2} \times 14.0) - 81.4\} \text{ MN m}^{-2}$$

$$= -6.4 \text{ MN m}^{-2}$$

$$\tau_{xy} = \{(\tfrac{1}{2} \times 135.4) - (\tfrac{1}{2} \times 14.0)\} \text{ MN m}^{-2}$$

$$= 60.7 \text{ MN m}^{-2}$$

For the $-45°$ ply

$$\sigma_x = 156.4 \text{ MN m}^{-2}$$
$$\sigma_y = -6.4 \text{ MN m}^{-2}$$
$$\tau_{xy} = -61.0 \text{ MN m}^{-2}$$

3. The specification for the space shuttle manipulator arm, Figure 6.11, calls for a tip positioning accuracy of 38 mm. Fully extended, Figure 6.12(a), the overall length from longeron in the payload bay to the tip of the boom is 15.25 m. However, the tip deflection in this mode of deployment is not design critical. Why? The boom sections are of tubular carbon fiber laminate construction. The mechanical properties are $E_{11} = 280$ GN m^{-2}, $E_{22} = 6.5$ GN m^{-2}, $\nu_{12} = 0.3$, $G_{12} = 6.5$ GN m^{-2} (after K. T. Kedward, UCLA, 1985).

The tip deflection of a weightless beam of length l by an end-load W is

$$\delta = \frac{Wl^3}{3EI}$$

where E is Young's modulus, and I is the moment of inertia of the cross-section about an axis through its center at right angles to the plane of bending [refer to question 10 in Section 1.6, and to J. H. Poynting and J. J. Thomson, *A Text-Book of Physics*, Charles Griffin, London, p. 91 (1907)].

So the tip deflection in the fully extended deployment mode (a) in Figure 6.12, is

$$\delta_{(a)} = \frac{Wl^3}{3EI}$$

In deployment mode (b), the tip deflection is

$$\delta_{(b)} = \frac{2W(l/2)^3}{3EI}$$

$$= \frac{Wl^3}{12EI}$$

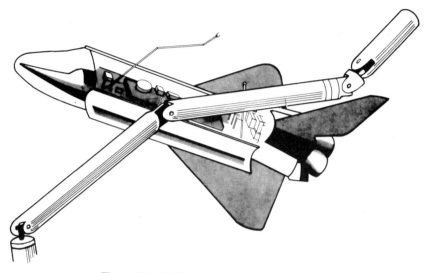

Figure 6.11. NASA's space shuttle manipulator arm.

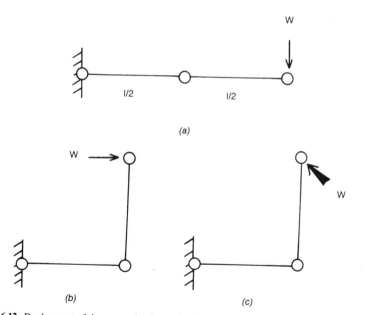

Figure 6.12. Deployment of the space shuttle manipulator arm. (a) Arms fully extended. Both arms loaded in bending. (b) Forearm at right angles to upper arm. Both arms loaded in bending. (c) Forearm at right angles to upper arm. Forearm loaded in bending, upper arm loaded in torsion.

In deployment mode (c), the tip deflection due to bending in the forearm is

$$\delta'_{(c)} = \frac{W(l/2)^3}{3EI}$$

to which must be added the tip displacement arising from torsion in the upper arm

$$\delta''_{(c)} = \left(\frac{l}{2}\right) \phi$$

where ϕ is the angular displacement in the upper arm. The couple producing this angular displacement is

$$W\left(\frac{\ell}{2}\right) = \frac{G \pi \phi}{2\left(\frac{l}{2}\right)} (r_1^4 - r_2^4)$$

where r_1 and r_2 respectively are the outer and inner radii of the upper arm. [Refer to F. H. Newman and V. H. L. Searle, *The General Properties of Matter*, Edward Arnold, London, p. 109 (1957).]

Hence

$$\delta''_{(c)} = \frac{Wl^3}{4\pi G(r_1^4 - r_2^4)}$$

Since

$$I = \int_{r_1}^{r_2} 2\pi r \, . \, dr \, . \, r^2$$

$$= \frac{1}{2} \pi(r_1^4 - r_2^4),$$

$$\delta''_{(c)} = \frac{Wl^3}{8GI}$$

So the overall tip deflection in deployment mode (c) is

$$\delta_{(c)} = \frac{Wl^3}{8EI} \left\{\frac{1}{3} + \frac{E}{G}\right\}$$

For an isotropic material $E = 2G(1 + \nu)$, Equation (6.3), and, taking

$\nu = 0.3$, then $E/G = 2.6$ and $\delta_{(c)} = 0.32\ Wl^3/EI$ which is marginally smaller than $\delta_{(a)} = 0.33\ Wl^3/EI$. For the laminate, $E_{11}/G_{12} = 42.8$, and $\delta_{(c)} = 5.4$ Wl^3/EI which is some 18 times larger than $\delta_{(a)}$. It is evident that the tip deflection in mode (c), and not that in mode (a), is the more critical.

6.5 Laminate code

Each ply is identified by the angle (ply angle θ), measured in degrees, between its fiber direction and the x-axis. The lamination sequence is written between square brackets, starting with the first ply laid down, i.e., starting at the lay-up tool surface. Adjacent parallel plies are shown by a subscript which denotes the number of such plies. Adjacent non-parallel plies are separated by an oblique line. An overall subscript T indicates that the total lamination sequence is shown. An overall subscript S is used to denote a symmetric laminate, in which case only one-half of the lamination sequence is listed. Adjacent plies of equal but opposite ply angle are denoted by $\pm\theta$. Positive ply angles are clockwise for an observer looking along the z-axis towards the origin, i.e., looking towards the lay-up tool surface. Symmetrical laminates with an odd number of plies are denoted by a bar over the counter ply. Quasi-symmetric laminates are denoted by overall subscript Q. Some examples are shown in Figure 6.13.

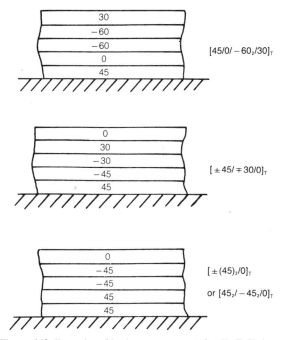

Figure 6.13. Examples of laminate sequences (after K. T. Kedward).

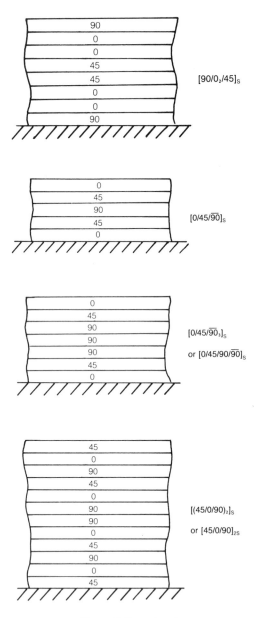

Figure 6.13. continued.

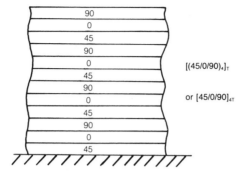

Figure 6.13. continued.

Sequences for which the fiber directions in adjacent plies are perpendicular to each other, for example [0,90] and [±45], are known as cross-plies. Non-perpendicular sequences are known as angle-plies.

6.6 Further examples

1. A certain material is made up from equal layers of two isotropic constituents in equal proportions by volume. The Young's modulus of the one is 10 times that of the other and they have the same Poisson's ratio. Derive simple approximate formulae for all the elastic constants of the composite assuming transverse isotropy about an axis normal to the plane of the layers. What will Young's modulus be at an angle of 45° to this axis (University of Bristol Examination, 1971)?

2. Compare the states of strain in a uniaxial composite subjected to pure shear (principal stresses equal and opposite) and bidimensional compression (principal stresses equal).

3. The mechanical properties of a unidirectional aramid fiber/epoxy resin composite include the following:

   - 0° tensile modulus 8.86 GN m^{-2}
   - 90° flexural modulus 0.6 GN m^{-2}
   - shear modulus 1.62 GN m^{-2}
   - major Poisson's ratio 0.435

   Write a computer program to solve the constitutive equation for uncoupled deformation of a symmetric laminate manufactured from this material as prepreg, and use it to determine the ply stresses in a [±60]$_s$ laminate subjected to the biaxial stress field $\sigma_x = 170$ MN m^{-2}, $\sigma_y = 40$ MN m^{-2}.

4. A two-ply laminate is made by adhesively bonding together 10 mm thick sheets of aluminum ($E = 70$ GN m^{-2}, $G = 26$ GN m^{-2}, $\nu = 0.33$) and steel

($E = 200$ GN m^{-2}, $G = 78$ GN m^{-2}, $\nu = 0.28$). Calculate the A_{ij}, B_{ij} and D_{ij} sub-matrices of the laminate constitutive equation, taking care to specify units. Comment on your answer for B_{ij}.

5. Physically, an n-ply laminate is a ($2n - 1$) layered material, of which n layers are fiber reinforced and $n - 1$ are neat resin. Examine the effect of the $n - 1$ layers of neat resin on the magnitude of Young's modulus measured by (a) in-plane uniaxial tension, and (b) bending, for $[0/45/90]_T$ and $[0/45/90]_s$ glassfiber/polyester laminates. For unidirectional laminate $E_1 = 40$ GN m^{-2}, $E_2 = 8.2$ GN m^{-2}, $G_{12} = 3.9$ GN m^{-2}, $\nu_{12} = 0.26$. Young's modulus for the neat polyester resin is 4 GN m^{-2}. The ply thickness is 0.1 mm. The neat resin layer thickness is 0.04 mm. What happens when the neat resin layers become vanishingly thin? Vis à vis heat sinks on glassfiber/resin laminate printed circuit boards, what would be the effect on Young's modulus measured by in-plane tension, of bonding a 0.5 mm thick aluminum film to the $[0/45/90]_T$ laminate? Young's modulus for aluminum is 70 GN m^{-2}.

6. Fiber reinforced laminate structures are exploited for use as pressure vessels, from automobile tires to storage tanks for hazardous fluids. What kind of experimental data on such structures would you collect for design purposes? In the case of automobile tires, how would anisotropy of rubber under stress affect your measurements? Define the conditions for which one of the Poisson's ratios of a laminate can be negative and comment on the implications of this fact in the context of pressure vessel design.

7. According to Figure 5.9, see Section 5.2, two plies are sufficient to remove tension/bending coupling. Verify that (i) three is the minimum number of plies needed to remove tension/torsion coupling, and (ii) two is the minimum number that removes bending/torsion coupling.

8. Sketch the laminate sequences $[0_2/90_2]_s$ and $[0/90]_{3s}$.

7

Anisotropy of thermal expansion

When the temperature of a laminate is changed, the resulting deformation may be specified by the strain tensor $[\epsilon_{ij}]$. Provided the temperature change ΔT is small and takes place uniformly throughout the laminate,* the deformation is homogeneous and all the components of $[\epsilon_{ij}]$ are proportional to ΔT

$$\epsilon_{ij} = \alpha_{ij}\Delta T \tag{7.1}$$

The coefficients α_{ij} are constants, the coefficients of thermal expansion.
As pointed out in Section 3.11, the ϵ_{ij} generated by thermal expansion constitute a matter tensor, that is, they are related to the symmetry of the laminate.
Since $[\epsilon_{ij}]$ is a symmetrical second rank tensor, $[\alpha_{ij}]$ is one also.

7.1 Thermal expansion of a 0° laminate

As noted by Horig[21] (see Chapter 5), the symmetry of a 0° laminate is that of an orthorhombic crystal; moreover, it is that of an orthorhombic holosymmetric crystal, the characteristic symmetry elements for which are three planes of symmetry intersecting in three mutually perpendicular diad axes. See Figure 7.1.
Consider the effects of this symmetry on

$$[\alpha_{ij}] = \begin{bmatrix} \alpha_{11} & \alpha_{12} & \alpha_{13} \\ & \alpha_{22} & \alpha_{23} \\ & & \alpha_{33} \end{bmatrix} \tag{7.2}$$

Figure 7.2 shows the effects of the diad parallel to axis Ox_3 on each of the components of the strain tensor.

---

*Here we ignore the effects of resin viscoelasticity in resin matrix composites and of metal plasticity in metal matrix composites, and treat thermal expansion as a problem in small strain elasticity.

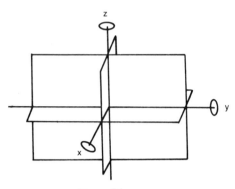

Figure 7.1.

There is no effect on ϵ_{11}, Figure 7.2(a), from which it is concluded that none of the components on the leading diagonal of the strain tensor is affected by the symmetry of the laminate. Figure 7.2(b) similarly shows that there is no effect on ϵ_{12}. For ϵ_{13}, however, operation of the diad along Ox_3 reverses the sign of the shear—see Figure 7.2(c). The same is true for ϵ_{23}.

Summarizing the effects of the diad parallel to Ox_3,

$$[\epsilon_{ij}] = \begin{bmatrix} \alpha_{11} \rightarrow \alpha_{11} & \alpha_{12} \rightarrow \alpha_{12} & \alpha_{13} \rightarrow -\alpha_{13} \\ & \alpha_{22} \rightarrow \alpha_{22} & \alpha_{23} \rightarrow -\alpha_{23} \\ & & \alpha_{33} \rightarrow \alpha_{33} \end{bmatrix} \Delta T$$

Since changes in sign are physically impossible, α_{13} and α_{23} must both be zero.

Referring to Figure 7.2(b), it is evident that the mirror plane perpendicular to axis Ox_1 (equivalent to an inverse diad parallel to Ox_1) causes a reversal of sign of ϵ_{12}, and hence of α_{12}. Since this is physically impossible, α_{12} must also be zero.

It is therefore concluded that a $0°$ laminate has only three coefficients of expansion.

$$[\alpha_{ij}] = \begin{bmatrix} \alpha_{11} & 0 & 0 \\ & \alpha_{22} & 0 \\ & & \alpha_{33} \end{bmatrix} \qquad (7.3)$$

Attempts have been made to relate the magnitudes of α_{11}, α_{22} (and α_{33}) to the thermal expansion coefficients and elastic constants of the fiber and matrix materials. For example, R. A. Schapery[25] finds for isotropic fibers in an isotropic matrix:

$$\alpha_{11} = \frac{E_f \alpha_f \eta + E_m \alpha_m (1 - \eta)}{E_f \eta + E_m (1 - \eta)} \qquad (7.4)$$

$$\alpha_{22} = (1 + \nu_m)\alpha_m(1 - \eta) + (1 + \nu_f)\alpha_f \eta - \alpha_{11}\{\nu_f \eta + \nu_m(1 - \eta)\} \qquad (7.5)$$

[25]R. S. Schapery. "Thermal Expansion Coefficients of Composite Materials Based on Energy Principles," *J. Composite Materials*, 2:380–404 (1968).

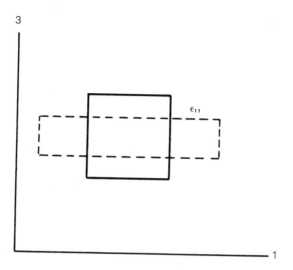

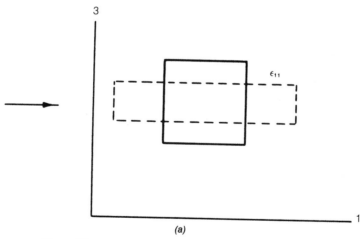

(a)

Figure 7.2. (a) ϵ_{22}, ϵ_{33} similar to ϵ_{11}, (b) ϵ_{12}, (c) ϵ_{23} similar to ϵ_{13}.

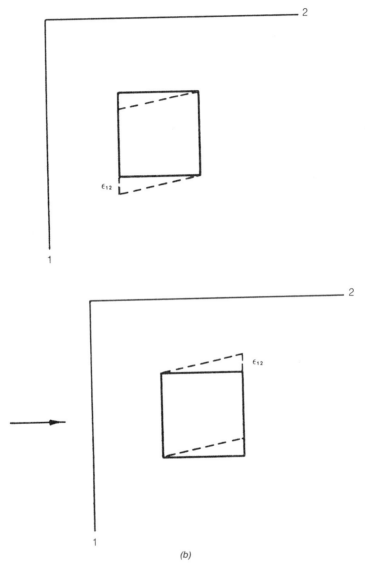

(b)

Figure 7.2. continued.

172

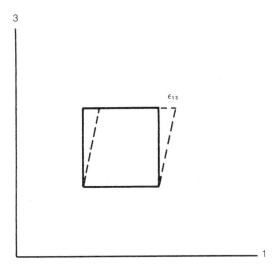

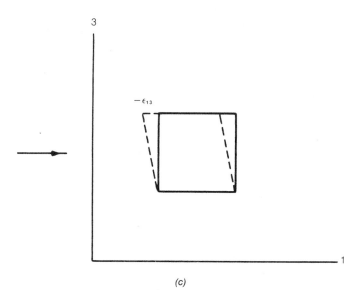

(c)

Figure 7.2. continued.

However, in practice it is usual to check the predicted α_{ij} by measurement. Typical values for a $0°$ glassfiber reinforced epoxy laminate with a fiber content $\eta = 1/2$ are

$$\alpha_{11} = 5.80 \times 10^{-6}\,K^{-1}$$

$$\alpha_{22} = 36.0 \times 10^{-6}\,K^{-1}$$

Typical measured values for a $0°$ high modulus carbon fiber reinforced epoxy laminate with a fiber loading $\eta = 3/5$ are

$$\alpha_{11} = -1.8 \times 10^{-6}\,K^{-1}$$

$$\alpha_{22} = 28.8 \times 10^{-6}\,K^{-1}$$

Note that $\alpha_{22} = 28.8 \times 10^{-6}\,K^{-1}$ is unacceptably large for some applications. Thus, for a $[0/90]_s$ laminate subjected to 300 K temperature variations between solar day and night, the 1% dimensional changes at right angles to each other would cause interlaminar cracks. To alleviate this possibility, it is usual to separate $0°$ and $90°$ plies with a $45°$ ply (or $\pm45°$ plies with a $90°$ ply).

Referring to the carpet plot shown in Figure 7.3 for coefficient of linear expansion in the laminate Ox-axis direction of $[0/\pm45/90]_s$ boron fiber/epoxy laminates, the lower, near horizontal line shows α_x for $[0/\pm45]_s$ laminates and the lines emanating from the lower line show the effects of replacing $\pm45°$s by $90°$s. By way of illustration of the effect of addition of just a few $\pm45°$ fibers to a $[0/90]_s$ laminate, Figure 7.3 reveals that $\alpha_x = 15.6 \times 10^{-6}\,K^{-1}$ for a 90% $0°$s/10% $90°$s lay-up, falling to $\alpha_x = 9.9 \times 10^{-6}\,K^{-1}$ for a 90% $0°$s/9% $90°$s/1% $\pm45°$s lay-up, and to $\alpha_x = 4.0 \times 10^{-6}\,K^{-1}$ for a 90% $0°$s/10% $\pm45°$s lay-up.

7.2 Warping of laminates

To inhibit warping in laminated plates during thermal expansion, it is necessary that the laminate sequence be symmetrical and balanced. Consider the following two laminate sequences, both of which are symmetrical about the mid-plane:

Ply number	1	2	3	4	5	6	7	8
Ply angle	$90°$	$+\theta$	$0°$	$-\theta$	$-\theta$	$0°$	$+\theta$	$90°$

Ply number	1	2	3	4	5	6	7	8	9	10
Ply angle	$90°$	$+\theta$	$+\theta$	$0°$	$-\theta$	$-\theta$	$0°$	$+\theta$	$+\theta$	$90°$

The second laminate contains unequal numbers of $\pm\theta$ plies, four $+\theta$ but only two $-\theta$ plies. Although it will not warp, when loaded in tension parallel to the $0°$ direction, the deficiency of $-\theta$ plies will allow shear displacement to occur during thermal expansion.

% 0° PLIES

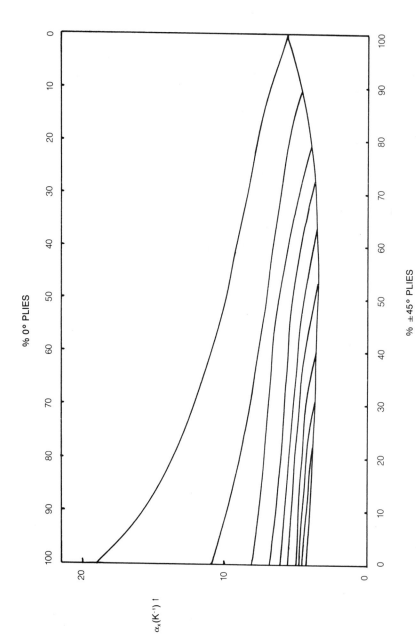

$\alpha_x (K^{-1})$ ↑

% ±45° PLIES

Figure 7.3. Linear coefficient of thermal expansion α_x for $[0/\pm45/90]_S$ boron fiber/epoxy laminates.

7.3 Dimensional stability of carbon fiber/epoxy resin laminates

It is of some technological interest to examine the requirements for a surface S of zero thermal expansion since such surfaces exist in carbon fiber laminates. Following F. C. Frank,[26] consider within the laminate, a sphere of unit radius:

$$x^2 + y^2 + z^2 = 1 \qquad (7.6)$$

During thermal expansion of the laminate, it expands and transforms to an ellipsoid:

$$\frac{x^2}{a^2} + \frac{y^2}{b^2} + \frac{z^2}{c^2} = 1$$

which, if the principal strains are ϵ_1, ϵ_2, ϵ_3 so that $a = (1 + \epsilon_1)$, $b = (1 + \epsilon_2)$, $c = (1 + \epsilon_3)$, can be rewritten

$$\frac{x^2}{(1 + \epsilon_1)^2} + \frac{y^2}{(1 + \epsilon_2)^2} + \frac{z^2}{(1 + \epsilon_3)^2} = 1 \qquad (7.7)$$

The lines in the sphere that do not change their lengths after deformation are the lines from the origin to points on the nonplanar closed loops of intersection* of the ellipsoid with the sphere—see Figure 7.4(a).

A condition for the existence of any such lines is that two of the principal strains have opposite signs, as is the case for the deformation created by thermal expansion of graphite fiber laminates (see Section 7.1). Any linear combination of Equations (7.6) and (7.7) defines a quadric surface passing through the line of intersection of the sphere and the ellipsoid, and for the particular combination

$$x^2 \left\{ \frac{1}{(1 + \epsilon_1)^2} - 1 \right\} + y^2 \left\{ \frac{1}{(1 + \epsilon_2)^2} - 1 \right\} + z^2 \left\{ \frac{1}{(1 + \epsilon_3)^2} - 1 \right\} = 0 \qquad (7.8)$$

this surface becomes, in general a quadric with apex at the origin. The generators of this quadric cone are the required directions of unchanged length.

---

[26]F. C. Frank. "On Dilatancy in Relation to Seismic Sources," *Rev. Geophys.*, 3:485–503 (1965); *ibid.*, 4:405–408 (1966).

*It is assumed that the expansion is small enough for the ellipsoid of transformation not to enclose the sphere.

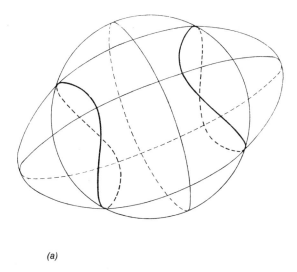

(a)

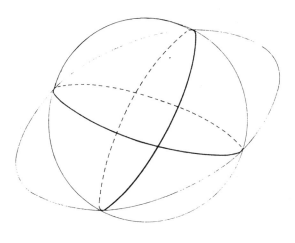

(b)

Figure 7.4. (a) Illustrating the general case of non-planar loops of intersection between a sphere and the ellipsoid to which it deforms. (b) The special case where one principal strain is zero.

In general, no part of this conical surface is plane. Suppose, however, that one of the principal strains, ϵ_3 say, is zero. Equation (7.8) becomes

$$x^2 \left(\frac{1}{(1 + \epsilon_1)^2} - 1 \right) + y^2 \left(\frac{1}{(1 + \epsilon_2)^2} - 1 \right) = 0$$

$$\frac{y^2}{x^2} = - \frac{\left(\dfrac{1}{(1 + \epsilon_1)^2} - 1 \right)}{\left(\dfrac{1}{(1 + \epsilon_2)^2} - 1 \right)} = \frac{\left(\dfrac{1 - (1 + \epsilon_1)^2}{(1 + \epsilon_1)^2} \right)}{\left(\dfrac{1 - (1 + \epsilon_2)^2}{(1 + \epsilon_2)^2} \right)}$$

$$= \frac{(1 + \epsilon_2)^2}{(1 + \epsilon_1)^2} \left(\frac{1 - (1 + \epsilon_1)^2}{1 - (1 + \epsilon_2)^2} \right)$$

$$= \frac{(1 + \epsilon_2)^2}{(1 + \epsilon_1)^2} \left(\frac{1 - 1 - \epsilon_1^2 - 2\epsilon_1}{1 - 1 - \epsilon_2^2 - 2\epsilon_2} \right)$$

$$= \frac{(1 + \epsilon_2)^2}{(1 + \epsilon_1)^2} \left(- \frac{\epsilon_1^2 + 2\epsilon_1}{\epsilon_2^2 + 2\epsilon_2} \right)$$

$$= \frac{(1 + \epsilon_2)^2}{(1 + \epsilon_1)^2} \left(- \frac{\epsilon_1}{\epsilon_2} \cdot \frac{2 + \epsilon_1}{2 + \epsilon_2} \right)$$

Hence

$$\frac{y}{x} = \pm \frac{1 + \epsilon_2}{1 + \epsilon_1} \left(- \frac{\epsilon_1}{\epsilon_2} \cdot \frac{2 + \epsilon_1}{2 + \epsilon_2} \right)^{1/2} \tag{7.9}$$

This is the equation of two planes, see Figure 7.4(b). That is, in the special case of one principal strain equal to zero, the undeformed surface is planar, in fact it is either of two planes. If the two planes are also unrotated, they satisfy all the requirements for S.

Thus either of the two planes, but not both simultaneously, can serve as S. If either plane is unrotated, the other rotates. Hence, the necessary and sufficient conditions for planar dimensional stability during thermal expansion are that the expansion is small, that the product of two of the principal strains be negative,

$$\epsilon_1 \epsilon_2 \leq 0 \tag{7.10}$$

and that the third principal strain be zero

$$\epsilon_3 = 0 \tag{7.11}$$

The plane must be a plane in one of the two special orientations relative to the state of strain, satisfying Equation (7.9). The state of strain is one in which all planes parallel to S before deformation are still parallel after deformation. The strain is thus conveniently described as compounded from a shear parallel to S and an expansion normal to S. This description is particularly convenient with relation to the intermediate stages of strain, during which the principal axes of strain rotate and the second undeformed plane not only rotates in space but is a continually changing plane in the material. Although at any stage in the deformation there are undeformed planes in two different orientations, planes in at most one orientation can remain undeformed through the deformation process.

Lightweight structural materials that have lines and, better still, planes of zero thermal expansion are of considerable importance for the manufacture of antennae used in outer space.

7.4 Thermal cycling of glassfiber/epoxy resin printed circuit boards

Paper- and fabric-based phenolic resin laminates were among the earliest materials used for electrical insulation (terminal boards, valve-holders, coil-formers, canes and rotating parts in switchgear). Most modern printed circuit boards are fabricated from glassfiber/epoxy resin laminates. The reasons for this include low dielectric constant and low cost of glassfiber laminates. A thin copper film is bonded to one or both sides of the laminate, photographically printed over with chemically-resistant ink, and the unprotected areas are removed by etching in an acid bath, leaving the required copper circuits bonded permanently to the laminate. There is one major problem with these boards, arising from thermal expansion mismatch between the laminate and the conductors and packages, deposited and/or mounted on the board. Soldering is carried out at temperatures as high as 260°C, and many solder joints do not survive the deformation induced by differential expansion and contraction between laminate and electronic package during heating from and cooling to ambient temperature. With the advent of multi-layer circuit boards, the provision of conducting paths from one layer to the next via copper plated through holes has introduced an in-service problem of metal fatigue and hence broken circuits in the plated through holes. Despite generous provision of heat sinks and forced cooling, temperature excursions of several tens of kelvins accompany on/off switching.

One successful method of measuring a lower limit on the magnitude of the cyclic stress experienced by plated through holes during temperature cycling is to bond to it a flexible membrane, accurately measure the displacement during flexing of the membrane and use Equation (6.22) to determine the stress acting normal to the membrane. Using a microscope cover slip as membrane, an optical flat is brought into close proximity to its free surface so that, when illuminated with monochromatic light (wavelength λ), a pattern of interference fringes (Fizeau fringes) is created in the space between cover slip and optical flat. These fringes relate precisely to the local thickness of the air gap between cover slip and optical

1 mm

Figure 7.5. Development of the displacement field around a plated-through hole in a multi-layer circuit board during a 3 s soldering operation (after D. A. Tossell).

flat, this thickness differing by $\lambda/2$ from one dark or bright fringe to the next.
When the cover slip flexes, as happens when the temperature of the underlying
laminate containing a copper plated through hole is caused to change, the pattern
of interference fringes changes. By superimposing on the pattern photographed
at the outset, successive images of the pattern photographed during temperature
cycling, Moire fringes faithfully representing changes in shape of the cover slip
are obtained. Figure 7.5 shows some examples. The Moire fringes are closed
loops concentric with the plated through hole. Successive Moire fringes delineate
loci of points that differ by half a wavelength in displacement normal to the cover
slip.

Rewriting Equation (6.22) the normal stress, that is the pressure p, giving rise
to the normal displacement w of the cover slip is given by

$$D\nabla^4 w = -p$$

where the flexural rigidity is

$$D = \frac{2Eh^3}{3(1 - \nu)}$$

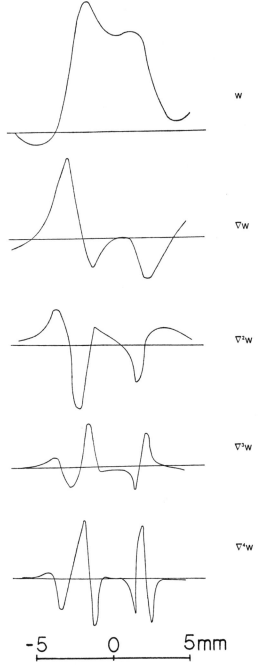

w

∇w

∇²w

∇³w

∇⁴w

-5   0   5mm

Figure 7.6. Normal displacement w across a diameter of the pattern of interference fringes seen in Figure 7.5, and the first four differentials of w.

$2h$ = cover slip thickness
E = Young's modulus of glass
ν = Poisson's ratio of glass

Figure 7.6 shows the normal displacement w, and its derivatives, across a diameter of the cover slip photographed in Figure 7.5. The normal stress is a maximum close to the boundary between the plated through hole and the laminate, it is compressive and is of the order of 3 MN m^{-2}. The expected fatigue lifetime for oxygen free high conductivity copper subjected to a stress amplitude of 3 MN m^{-2} is 10^4 cycles. Note also the large normal tension beyond the hole, that is, within the laminate. This tension is responsible for delaminations and for debonding of conductor strips in thermally cycled circuit boards.

7.5 Thermal expansion mismatch in metal matrix and ceramic matrix composites

Failure of short fiber reinforced metals is usually by the coalescence of voids that have nucleated at fiber ends. It is asking a lot of the strength of adhesion between fiber and matrix to expect the fiber ends to remain bonded to the matrix material after the differential displacement that occurs during mechanical deformation. The same is true of the differential axial displacement due to any difference in thermal expansion coefficients between the fiber and matrix materials and, in both cases, the consequence for metal matrix composites is void nucleation.

Ceramic matrix composites, on the other hand, usually fail by matrix cracking and, in the case of short fiber reinforced ceramics, the cracks often propagate from the fiber ends.

For both metal matrix and ceramic matrix composites, it is evident that close matching of coefficient of thermal expansion for fiber and matrix materials is an important consideration if temperature cycling is envisaged. Tables 7.1, 7.2 and 7.3 give the coefficients of linear expansion (α) for a selection of matrix materials. Table 7.4 gives values of α for some common fibers.

7.6 Further examples

1. A square ABCD, of measured size, is scribed on one face of a laminate. It is heated by ΔT so that thermal expansion takes place and the displacements, in the plane of the face, of the points A, B, and C are carefully measured. How would you calculate (i) the components of the strain tensor, referred to suitable axes, (ii) the volume coefficient of thermal expansion, (iii) the stretch of the diagonal AC, (iv) the stretch of the other diagonal BD? Make clear at what point in your answer you make use of the symmetry of the laminate.

Table 7.1. Coefficients of linear expansion for candidate polymer matrix materials (units of 10^{-6} K^{-1}).

Polyester resins	70–110
Epoxy resins	50–80

Table 7.2. Coefficients of linear expansion for candidate metal matrix materials (units of 10^{-6} K^{-1}).

36% Ni-Fe (invar)	0–1.5
Zirconium (Zr)	5.8
Titanium	8.4
Nodular cast iron	12.4
Nickel	13.3
Aluminum (Al)	23
4% Cu-Al (duralumin)	27

Table 7.3. Coefficients of linear expansion for candidate ceramic matrix materials (units of 10^{-6} K^{-1}).

Silicon nitride (Si_3Ni_4)	2.6
Silicon carbide (SiC)	4.4
Graphite (C)	4.5
Alumina (Al_2O_3)	7.7
Partially stabilized zirconia (ZrO_2)	10.6

Table 7.4. Coefficients of expansion for a selection of fibers (units of 10^{-6} K^{-1}).

S glass fiber	5.6
E glass fiber	4.9
High modulus carbon fiber	0.9 axial, 10.8 radial
Aramid fiber	2 axial, 59 radial
Boron fiber	4.9
Silicon carbide	3.1

2. A 40 mm diameter pultruded carbon fiber/epoxy resin rod, 20 m long at 18°C, is inserted into the gap between rigid supports 20 m apart. What is the compressive thrust in the rod if its temperature is lowered to $-35°C$? The axial coefficient of linear expansion is -1.5×10^{-6} K^{-1}, and the axial Young's modulus is 200 GN m^{-2}.
3. Discuss the thermal expansion and contraction of a carbon fiber, filament wound spherical shell (see Chapter 2, further Example #4).

8

Fracture and fracture
mechanics

"...toughness means possession of some
physical mechanism(s) for rendering cracks
non-disastrous..." F. C. Frank[1]

Before 1921, there were two commonly held hypotheses for the fracture of brittle solids, that is, of solids whose deformation is elastic up to the point of failure. Rupture was expected if either the maximum tensile stress or the maximum extension exceeds a critical value. By that time the destructive influence of surface scratches was also a matter of common knowledge. Indeed this influence had long since been turned to useful account in the process of glass cutting; a fine scratch scribed on the surface of plate glass produces such a local weakness to tension that a crack along the line of the scratch can be propagated by applied forces which produce insignificant stresses in the rest of the plate. Now a crack is physically indistinguishable from a narrow elliptic hole and Inglis had shown that the stress around an elliptic hole depends on the shape of the ellipse, Equation (4.4), and not on its absolute size. The predicted insensitivity of maximum tensile stress (or of maximum extension) to crack size worried Griffith since it appears to contradict the observation that reducing the size of surface scratches by polishing increases the strength of solids. Therefore, Griffith looked for a new fracture hypothesis.

8.1 Griffith's theory

On the basis that the equilibrium state of an elastically deformed solid is such that the overall work done is a minimum, A. A. Griffith[7] argued that, if the system can pass from the unbroken to the broken condition by a process involving a continuous decrease in the overall work done, then the equilibrium state must be one in which fracture has occurred.

To put the theory on a quantitative basis, consider a plate subjected to constant uniaxial tensile forces applied at its edge – Figure 8.1(a).

Using Inglis's calculation, the strain energy of the material within the ellipse α,

185

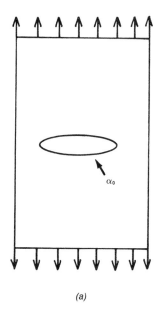

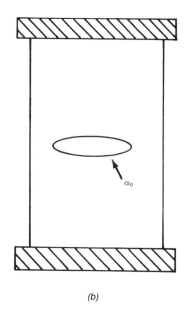

(a)   (b)

Figure 8.1.

per unit thickness of stressed plate, is

$$\frac{1}{2}\int_0^{2\pi} \frac{u_\alpha}{h} \cdot \sigma_{\alpha\alpha} d\beta + \frac{1}{2}\int_0^{2\pi} \frac{u_\beta}{h} \cdot \sigma_{\alpha\beta} d\beta \tag{8.1}$$

(Note that $\sigma_{\beta\beta}$ merely moves energy around inside the ellipse.)

When an elliptic hole $\alpha = \alpha_0$ is introduced, the strain energy increases and, at large α, tends toward the value

$$\frac{\pi c^2 \sigma^2}{8\mu}\left\{\frac{1}{2}(p-1)e^{2\alpha} + (p+1)\cosh 2\alpha_0\right\} \tag{8.2}$$

i.e., the plate *gains* a quantity of strain energy

$$\Delta W_{internal} = \frac{\pi c^2 \sigma^2}{8\mu}(p+1)\cosh 2\sigma_0 \tag{8.3}$$

In the limiting case of a very narrow ellipse $\alpha_0 = 0$, $\cosh 0 = 1$ and

$$\Delta W_{internal} = \frac{(p+1)\pi c^2 \sigma^2}{8\mu} \tag{8.4}$$

($\alpha_0 = 0$ is a good approximation to an internal crack and $2c$, the length of the focal line, is very nearly the crack length.)

The increase in strain energy is a consequence of external work done by the constant forces, and to calculate the latter, Griffith makes use of the following theorem concerning the strain energy of a Hookean body. If such a body is deformed from the unstrained state to equilibrium by constant surface forces, its strain energy is equal to half the work done by the external forces, the rest of the external work being dissipated as heat, etc. A rigorous proof of this theorem is given in A. E. H. Love,[12] p. 173. Its physical plausibility is demonstrated by consideration of a dead load tensile test. If Hooke's law is obeyed, the internal stress builds up linearly with strain. The strain energy of the deformed body is given by the shaded area, $(1/2) F \Delta l$ in Figure 8.2. However, the work done by F acting through the displacement Δl from the unstressed state to the state of equilibrium is $F \Delta l$. Thus

$$\Delta W_{external} = -2 \Delta W_{internal} \qquad (8.5)$$

According to Griffith's postulate, fracture occurs if

$$\frac{\partial}{\partial c} \Delta W_{overall} = 0 \qquad (8.6)$$

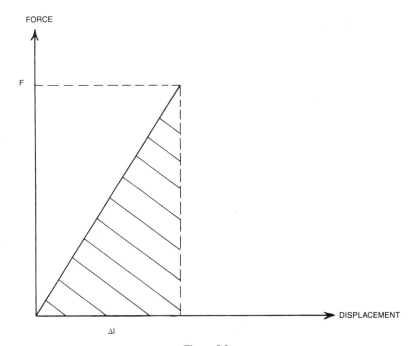

FORCE

F

DISPLACEMENT

Δl

Figure 8.2.

$$\Delta W_{overall} = \Delta W_{external} + \Delta W_{internal} + \Delta U \tag{8.7}$$

ΔU, the surface energy of the crack per unit thickness of plate, is given by

$$\Delta U = -4c\gamma \tag{8.8}$$

where γ is the specific fracture surface energy, that is the energy per unit area of fracture surface. Thus

$$\Delta W_{overall} = \frac{(p + 1)\pi c^2 \sigma^2}{8\mu} - 4c\gamma \tag{8.9}$$

and the condition for fracture is that

$$(p + 1)\pi c \sigma^2 = 16\mu\gamma \tag{8.10}$$

Now consider the case of fracture in a plate tensioned between fixed clamps, Figure 8.1(b). Introduction of a sharp internal crack of length $2c$ causes the plate to lose a quantity of strain energy

$$\Delta W_{internal} = \frac{(p + 1)\pi c^2 \sigma^2}{8\mu} \tag{8.11}$$

(It is odd that the loss of internal energy that occurs when a stretched membrane is slit should be equal in magnitude to the internal energy gained when the same membrane is fractured by the action of constant external forces. The identity is a consequence of the mathematics which describe the two modes of fracture.) Since the external forces do not move during fracture in a plate tensioned between fixed grips

$$\Delta W_{external} = 0 \tag{8.12}$$

Hence

$$\Delta W_{overall} = \frac{(p + 1)\pi c^2 \sigma^2}{8\mu} - 4c\gamma \tag{8.13}$$

and the condition for fracture is again

$$(p + 1)\pi c \sigma^2 = 16\,\mu\gamma \tag{8.14}$$

It is therefore concluded that this is the general condition for fracture. For deformations approximating to plane strain (one principal strain equal to zero),

$$p = 3 - 4\nu$$

Table 8.1. Critical Griffith crack lengths calculated from Equation (8.16).

Material	Young's Modulus E(GN m^{-2})	Specific Fracture Surface Energy γ (J m^{-2})	Tensile Strength (MN m^{-2})	Critical Griffith Crack Length c (m)
Polyester resin	2–4.5	50–500 (200)	40–90	$(0.2–4) \times 10^{-7}$
Epoxy resin	3–6	50–500 (200)	35–100	$(0.2–7.8) \times 10^{-7}$
E glass fiber	68–76	0.5–5	3.5×10^3	$(2–17) \times 10^{-12}$
Type I carbon fiber	390	41–49*	2.2×10^3	$(2.1–2.5) \times 10^{-9}$
Type II carbon fiber	250	44–56*	2.7×10^3	$(0.96–1.2) \times 10^{-9}$
Kevlar	130	50**	3.6×10^3	3.2×10^{-10}

*Specific surface free energy, not specific fracture surface energy.
**Assumed value.

and for deformations approximating to plane stress (one principal stress equal to zero),

$$p = \frac{3 - \nu}{1 + \nu}$$

Substituting for

$$\mu = \frac{E}{2(1 + \nu)},$$

the fracture stress is

$$\sigma = \sqrt{\frac{2E\gamma}{(1 - \nu^2)\pi c}} \quad \text{plane strain} \tag{8.15}$$

$$\sigma = \sqrt{\frac{2E\gamma}{\pi c}} \quad \text{plane stress} \tag{8.16}$$

So Griffith's concept of fracture is that solids contain large numbers of small cracks oriented at random and that, under the influence of externally applied uniaxial tension, the longest crack oriented perpendicular to the tensile axis propagates when the tensile stress at its edge reaches a critical value. The distribution of pre-existing cracks in real materials and its effect on strength was first considered by W. Weibull[27] and has led to a branch of fracture theory known as sta-

[27]W. Weibull. "A Statistical Theory of the Strength of Materials," *Ingeniors Vetenskaps Akademien*, Handlingar Nr 151, pp. 1–51 (1939).

tistical fracture mechanics, see Section 8.2. Note that Equations (8.15) and (8.16) for fracture strength are independent of crack shape, i.e., Griffith's theory explains why polishing increases the strength of solids. Also, since the longest favorably oriented crack likely to be present becomes progressively smaller with smaller test pieces, Griffith's theory explains why the fracture strength of fibers increases with decreasing fiber diameter—Figure 1.4.

In Table 8.1 the measured fracture stresses of several materials used to fabricate fiber composites have been substituted into Equation (8.16) to estimate the critical Griffith crack length, that is the length of the pre-existing crack responsible for failure if the deformation is wholly elastic up to failure.

8.2 Statistical aspects of fracture

According to Griffith's theory, brittle materials, and especially the free surfaces of brittle materials, are peppered with a distribution of flaws any of which, if suitably stressed, can lead to catastrophic failure of the material. Each flaw can be characterized by its length or by the critical stress at which it propagates. If an increasing tensile stress is applied to the material, fracture will occur when the

Figure 8.3. Mathematical model for a uniaxial composite (see text).

applied stress reaches the critical stress for the most dangerously oriented flaw. Thus, if a bundle of resin bonded fibers be regarded as comprised of a large number of units, each containing one flaw, the strength of the solid will be determined by the strength of the weakest unit, i.e., that containing the most severe flaw. This weakest link concept forms the basis of Weibull[27] statistics for the fracture of brittle solids. The division of the composite into units, that is, into a chain of short bundles in series, is illustrated in Figure 8.3.

The flaws in a given unit can be considered as part of an infinite population of such flaws characterized by some distribution function $F(x)$. Here x can be considered either as a flaw length or as the critical tensile stress at which a flaw propagates; in practice it is usually more convenient to define x as the latter. The problem is to determine the minimum critical tensile stress, or maximum effective pre-existing crack length, which will occur in a sample of known size drawn from such a population. Since we are dealing with a sample drawn from an infinite population we cannot determine the information precisely but must use concepts of probability. This problem is considered in a branch of statistical theory termed the *statistics of extremes* since it deals with the probability distribution of extreme values in samples drawn from an infinite population having a known distribution.

The cumulative distribution function, $F(\sigma)$, can be defined as follows. Let $P(\sigma \leq \sigma')$ be the probability of selecting at random an individual flaw from the population with a critical stress less than or equal to σ'. Then

$$P(\sigma \leq \sigma') = F(\sigma') \tag{8.17}$$

We can also define a probability density function $f(\sigma)$ given by

$$f(\sigma) = \frac{dF(\sigma)}{d(\sigma)} \tag{8.18}$$

Again

$$P(\sigma') = f(\sigma') \tag{8.19}$$

where $P(\sigma')$ is the probability of selecting a flaw from the population with a critical stress equal to σ'. In a sample of N flaws, the most probable number having $\sigma = \sigma'$ will be $Nf(\sigma')$ and the most probable number having a $\sigma \leq \sigma'$ will be $NF(\sigma')$.

By the weakest link principle, the probability that a given specimen will fail at a stress σ' can be equated with the probability that the smallest critical fracture stress in the sample of flaws included in the specimen is equal to σ'. For a sample of N flaws drawn from a population characterized by a cumulative distribution function $F(\sigma)$, the probability $g(\sigma)$ that any critical stress σ is a least value can be shown to be

$$g(\sigma) = N\{1 - F(\sigma)\}^{N-1} f(\sigma) \tag{8.20}$$

The associated cumulative probability function $G(\sigma)$ for the composite is obtained by integration of Equation (8.20)

$$G(\sigma) = \int_0^\sigma g(\sigma)d\sigma$$

$$= 1 - \{1 - F(\sigma)\}^N \tag{8.21}$$

The function $F(\sigma)$ in Equation (8.21) can be written in the form

$$F(\sigma) = 1 - \exp\{-\phi(\sigma)\}$$

where $\phi(\sigma)$ is often termed the risk of rupture and must be a positive, non-decreasing function of σ. Two suitable forms for this function were suggested by Weibull, namely

$$\phi(\sigma) = \left(\frac{\sigma}{\sigma_0}\right)^M \tag{8.22}$$

or

$$\phi(\sigma) = \left(\frac{\sigma - \sigma_u}{\sigma_0}\right)^M \tag{8.23}$$

Here σ_u, σ_0 and M are constants, characteristic of a particular surface flaw distribution.

Hence

$$G(\sigma) = 1 - \exp\{-\phi(\sigma)\} \tag{8.24}$$

and using the respective expressions for $\phi(\sigma)$ we obtain

$$G(\sigma) = 1 - \exp\left\{-N\left(\frac{\sigma}{\sigma_0}\right)^M\right\} \tag{8.25}$$

and

$$G(\sigma) = 1 - \exp\left\{-N\left(\frac{\sigma - \sigma_u}{\sigma_0}\right)^M\right\} \tag{8.26}$$

The difference between the above two functions is that in the former the probability of failure at stress σ tends to zero as σ tends to zero whilst in the latter, zero failure probability occurs at some higher stress level σ_u. The physical significance of non zero σ_u is that it corresponds to an upper limiting flaw size in the flaw distribution.

The inclusion of a limit parameter seems intuitively reasonable and it is claimed that the three parameter function better describes the experimental data for monolithic solids. In practice, the two parameter function is usually used to describe the strength of fibers and the three parameter function is used for monolithic solids.

One consequence of the mathematics of this weakest link model is that, when N is large, it is the extreme lower end of the distribution function $F(\sigma)$ that determines the form of $G(\sigma)$. To take account of this fact we first need to develop the model for failure of the composite.

When a fiber fractures, it no longer supports load in the short bundle (unit) that contains the fracture. (Further away from the fracture it will support load in accordance with shear lag theory—see Section 10.5.) Since the sum of the loads carried by the remaining load-bearing fibers must equal the overall load supported by the composite, it is necessary to prescribe a load sharing formula for the remaining fibers. The simplest formula is equal redistribution of the load formerly carried by the fiber that has failed. However, this formula pre-supposes that the remaining fibers are identically taut. In a real composite there will be a distribution of slack, in which case a more realistic formula might be that the first nearest neighbors bear most of the extra load, that the second nearest neighbors bear a smaller fraction, and that more distant neighbors bear smaller fractions still. To make calculation of $F(\sigma)$ tractable, D. G. Harlow and S. L. Phoenix[28] consider a two-dimensional bundle of fibers. When one fiber fails, they assume that the two fibers either side of it each become subjected to one and a half times their previous loads. If two adjacent fibers fail, the two fibers either side of the failed pair each become subjected to twice their previous loads, and so on. Hence, if r is the number of adjacent fibers to fail, each of the two fibers either side of the group that has failed becomes loaded to $(1 + r/2)$ times its previous load.

$$K = 1 + r/2 \tag{8.27}$$

is called the load concentration factor.

Identifying $A^{[2]}(\sigma)$ as the occurrence of two or more adjacent fiber failures, and $\overline{A}^{[2]}(\sigma)$ as the complementary event that no two or more broken fibers are adjacent to each other, the functions to be calculated are

$$F^{[2]}(\sigma) = \phi(\sigma)r\{A^{[2]}(\sigma)\} \tag{8.28}$$

[28]D. G. Harlow and S. L. Phoenix. "Probability Distributions for the Strength of Fibrous Materials Under Local Load Sharing I: Two-Load Failures and Edge Effects," *Adv. Appl. Prob.*, 14:68–94 (1982).

and

$$Q^{[2]}(\sigma) = \phi(\sigma)r\{\overline{A}^{[2]}(\sigma)\}$$
$$= 1 - F^{[2]}(\sigma)$$

(8.29)

$F^{[2]}(\sigma)$ is an upper bound for $F(\sigma)$ in Equation (8.21) since the occurrence of two or more adjacent fiber failures is a necessary but not sufficient criterion for overall bundle failure. By developing a recursive system of equations, Harlow and Phoenix obtain expressions for $F^{[2]}(\sigma)$ and go on to compute the failure probability curves shown in Figure 8.4.

Typical failure strains in composites for the common fibers are E glass fiber 0.024, S glass fiber 0.029, high strength carbon fiber 0.011, ultrahigh modulus carbon fiber 0.002, aramid fiber 0.018–0.028, boron fiber 0.007, silicon carbide fiber 0.006–0.012.

8.3 Fracture mechanics

Formally, the Griffith criterion, Equation (8.6), is

$$\frac{\partial}{\partial c}\left\{ \frac{-(p + 1)\pi c^2 \sigma^2}{8\mu} + 4c\gamma \right\} = 0$$

(8.30)

G. Irwin[29] pointed out that the first term in brackets on the left hand side of Equation (8.30) is the rate, with respect to crack extension (not with respect to time), of loss or gain of strain energy density. (Note that the appearance of a crack in a plate tensioned between fixed grips is accompanied by a loss of strain energy, whereas appearance of a crack in a plate subjected to constant applied tensile forces is accompanied by a gain of strain energy equal in magnitude to the former loss. That the two quantities of energy are equal is a consequence of the mathematics — see Section 8.1.) Thus, the Griffith criterion is described in linear elastic fracture mechanics notation by the equation

$$\mathscr{G} = \text{constant}$$

(8.31)

where $\mathscr{G}$, after Griffith, is the strain energy release rate. To be precise, for the problem considered by Griffith, namely fracture in a solid which undergoes only elastic deformation up to the point of failure,

$$\mathscr{G} = 2\gamma$$

(8.32)

where γ is the specific fracture surface energy.

---

[29]G. Irwin. "Fracturing of Metals," *ASM Symposium*, Chicago, pp. 147–166 (1947).

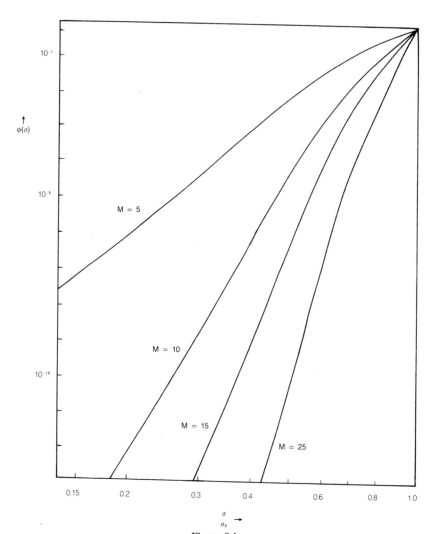

Figure 8.4.

Griffith's hypothesis tells us that the strain energy for conversion to fracture surface energy comes from the system as a whole; in fact, it mostly comes from afar because there is nowhere near enough energy available close to the crack. Fracture mechanics has exploited this concept of stress concentration created at a crack edge by stress applied at a distance. The consequent stress field near the crack edge and especially the singularity at the edge, is given by the differential (that is the slope) of the displacement field. In linear elastic fracture mechanics the crack profile and hence the displacement field near the crack are assumed to be parabolic, and not elliptic as assumed by Griffith. Thus

$$\Delta u = \frac{\mathcal{K}_I}{(2\pi)^{1/2}} \cdot r^{1/2} \cdot \frac{8(1 - \nu^2)}{E} \tag{8.33}$$

where Irwin's stress intensity factor $\mathcal{K}_I$ is given by the relationship

$$\mathcal{G}_I = \frac{\mathcal{K}_I^2}{E}(1 - \nu^2) \tag{8.34}$$

E is Young's modulus.

In a homogeneous mixture of fibers and matrix material, it is probably alright to use a generalized strain energy integral in order to evaluate $\mathcal{G}$ and hence $\mathcal{K}$. More care is needed when evaluating γ. Overall failure incurs fiber fracture, resin fracture, and fiber/resin interfacial failure. Surface energies of solids are of the order of 1 J m^{-2} so $\mathcal{G}_{critical}$, the strain energy release rate predicted by Equation (8.32) for an elastic solid, might also be of order 1 J m^{-2}. This is the case for ceramic materials including glass, diamond and boron. Interfacial energies, on the other hand, that is the energies of interfaces between solid phases, vary widely from 10^{-2} J m^{-1} for coherent twins in crystals to 1 J m^{-2} for grain boundaries. So some dilution in the magnitude of $\mathcal{G}_{critical}$ might be expected for multiphase materials, the fracture of which includes fracture at interfaces.

The critical values of $\mathcal{G}$ and $\mathcal{K}$ at which cracks propagate as Griffith cracks, that is catastrophically, are both used as measures of toughness.* When measured for cracks propagating perpendicular to the direction of largest tensile principal stress, these values are denoted $\mathcal{G}_{Ic}$ and $\mathcal{K}_{Ic}$ where subscript I denotes "mode I" (see below and Figure 8.9) and subscript c denotes "critical."

The relative orders of magnitude for $\mathcal{G}_{Ic}$ for different materials are 1 J m^{-2} for glass, 10 J m^{-2} for sintered ceramics, 10^4 J m^{-2} for light alloys, 10^2–10^3 J m^{-2} for fiber reinforced plastics, 10^4–10^5 J m^{-2} for fiber reinforced metals, and 10^5 J m^{-2} for some steels. The corresponding orders of magnitude for $\mathcal{K}_{Ic}$ are <1 MN

---

*Readers versed in dimensional analysis may sympathize with purists who object to the use of $\mathcal{K}$ as a measure of toughness. The SI unit for $\mathcal{K}$ is N m$^{-3/2}$ which contains the square root of a volume in the denominator. The use of strain energy release rate $\mathcal{G}$ as a measure of toughness has rather more physical meaning.

Figure 8.5a. Maxwell element.

m$^{-3/2}$ for glass, 1–10 MN m$^{-3/2}$ for ceramics, and 10–10^2 MN m$^{-3/2}$ for metals, for fiber reinforced plastics and for fiber reinforced metals.

Symmetric cross-plies [±45°]$_s$ exhibit large toughness and this is usually attributed to their large strain capacity. However, ±45° plies incur large values for major Poisson's ratio (see Chapter 6, Section 6.3), so cross-ply laminates distort easily, that is they are not very stiff. Hence their use is restricted mainly to shear webs. The next to toughest laminates are the pseudo-isotropic lay-ups, for example [0/ ±60]$_s$ and [0/ ±45/90]$_s$. For these the in-plane stiffnesses along the laminate axes are identical, but the bending stiffnesses are different. Another rule of thumb worthy of note is that thermoplastics are generally tougher than thermosets.

The question "*where* is the stored energy stored, that is released during fracture?" is usually answered as follows: materials with zero or infinite modulus are useless for storing strain energy. Strain energy density is given by the integral of the product of stress and strain, so zero stress (in a zero modulus material) or zero strain (in an infinite modulus material) are equally useless. Consider a composite in which fiber and matrix materials are subjected to the same stress. A chain of alternately high modulus and low modulus links, Figure 8.5a, is a mathematical model for such a composite. Most of the strain energy is in the low

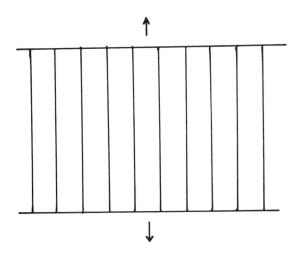

Figure 8.5b. Voight element.

modulus links. On the other hand, if we load a parallel array of steel and rubber wires, Figure 8.5b, most of the strain energy is in the steel.

In practical composites, it is usually reckoned to have a more or less equal division of strain energy between fiber and matrix materials. So, although most of the strain energy released during fracture of resin bonded fiberglass might be consumed by the resin, prior to the fracture it is stored about half in the fibers and half in the resin.

Whilst acknowledging the Griffith premise that the strain energy converted to fracture surface energy must come from afar, it is a fact that it is local forces which do the work of fracture. Fracture phenomena where local parameters are especially expected to be of overriding importance include delamination (see Chapter 9) and adhesive joint failures (see Chapter 11).

The subject of fracture is further complicated by the fact that it is possible for cracks to propagate at stresses below the critical Griffith fracture stress. This phenomenon, known as sub-critical crack growth, is prevalent during fatigue—that is during prolonged cyclic stressing to stress amplitudes that are well below the Griffith strength—during corrosion fatigue, and during stress cracking (stress corrosion), see question 3 in Section 8.9.

8.4 Origins of toughness

In practice, the strain energy released during crack growth is usually more than sufficient to account for creation of the crack surfaces. This is because most materials possess mechanisms, in addition to fracture, for relief of the stress concentration at the edge of a crack—plastic deformation is one mechanism. To take into account plastic work done during fracture, Orowan[30] rewrote the Griffith criterion as

$$\mathscr{G} = 2(\gamma + \gamma_{plasticity}) \qquad (8.35)$$

Possession of mechanisms other than fracture for relieving stress at crack edges is, of course, the origin of toughness.

One reason why plastics are tough has to do with the way stress is supported through tangles of chains, some of which can be straightened out. The number of configurations into which these chains can then change is lowered; that is, the configurational entropy is lowered by application of stress. If cross-links are present, more and more mechanical support comes from the free energy $F = U - TS$. The upshot is that the stress concentration at a sharp notch is less localized, say, in polyethylene than it is in diamond. This natural spreading of the stress is one reason for the toughness of rubber and plastics. Another has to do

---

[30]E. Orowan. "Fracture and Strength of Solids," *Reports on Progress in Physics, Vol. XII*, pp. 185–232 (1948–1949).

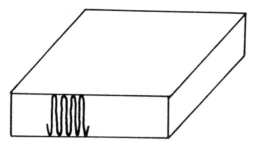

Figure 8.6. Illustrating a polymer lamella single crystal. In polyethylene, the fold length contains about 100 carbon atoms.

with the crystalline nature of some polymers. The way long molecules (length $\cong 1\mu m$) crystallize is to fold themselves up like a Chinese cracker, Figure 8.6. This folded-up molecule can actually be pulled out. Cracks bridged by parts of chains can be seen in the electron microscope. The origin of the bridging ropes involves the change from spaghetti-like configurations into folded chains that are then ready for pulling out. By this means, crystalline polymers can support cracks.[1]

In metals, toughness has its origins in the mechanisms of plastic yield. These mechanisms include dislocation generation, multiplication and entanglement, mechanical twinning and grain boundary sliding. In Chapter 4, Section 4.1, it was pointed out that Equation (4.5) predicts stress concentrations of the order of 100 at the edges of cracks which, for metals subjected to normal design loads, is far in excess of the yield stress. Hence the material ahead of a crack yields plastically and this yielded zone usually takes the form of two lobes oriented by about 45° to the axis of largest tensile principal stress as shown in Figure 8.7.

In some metals, the extent of the yielded zone can be revealed by etching; the yielded material contains a high dislocation density and is preferentially attacked by the etchant.

Irwin's method for taking into account the plastic work done during crack propagation is to include the amount of energy represented by the area, between σ_{yield} and $\sigma_{concentration}$, under the curve shown in Figure 4.2(b). This method gives, for the plastic zone size

$$r_y = \frac{1}{2\pi}\left(\frac{\mathscr{K}}{\sigma_y}\right)^2 \tag{8.36}$$

A more realistic measure of the extent of the plastic zone is

$$d_y = 2r_y = \frac{1}{\pi}\left(\frac{\mathscr{K}}{\sigma_y}\right)^2 \tag{8.37}$$

Other methods to take into account the plastic work done include (i) summation

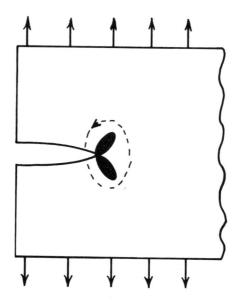

Figure 8.7. The yielded zone, ahead of a crack, responsible for the toughness of metals.

of the work done by point loads, per unit length of crack edge, of magnitude $\sigma_{yield}dt$ acting on columns of thickness dt – see Figure 8.8, and (ii) calculation of the work required to introduce equivalent dislocation displacement fields – see Figure 8.9. An imaginary array of edge dislocations with Burger's vector perpendicular to the crack could generate the mode I displacement field, and imaginary arrays of screw dislocations with Burger's vectors parallel to the crack could generate modes II and III displacement fields.

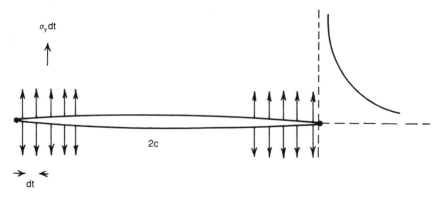

Figure 8.8. Dugdale model for the plastic zone as an array of columns each of thickness dt loaded by point loads of magnitude $\sigma_{yield} \cdot dt$.

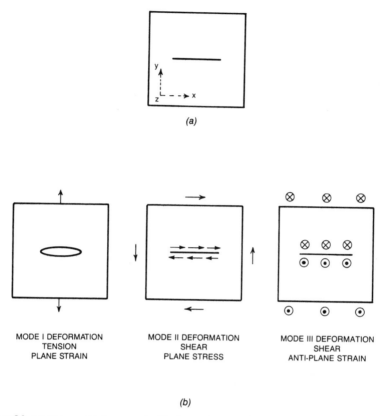

(b)

Figure 8.9. (a) Plate containing a crack, (b) the crack in (a) could be opened by any of the three systems of forces shown here.

Note that the occurrence of plastic deformation means that the critical Griffith crack length is larger than indicated in Table 8.1.

In fiber reinforced materials (plastics and metals) the region of plastic deformation ahead of the crack edge is usually referred to as the damage zone; the plastic zone in the matrix phase is strongly inhomogeneous, due to the presence of the elastically stronger fibers. The mechanisms of deformation include molecular orientation in the case of polymer matrix composites, and yield in the case of metal matrix composites. In both cases, and in the case of fiber reinforced ceramics, the bridging of cracks, by fibers only partially pulled-out, is a further source of toughness. This toughening in the wake of a crack is analogous to the contribution to the toughness of polymers by crack bridging chains discussed above. Figure 8.10, after P. W. R. Beaumont, illustrates the toughening roles played by fibers bridging a crack. Close to the crack edge, the crack opening displacement is small enough to be accommodated by enhanced extension of the

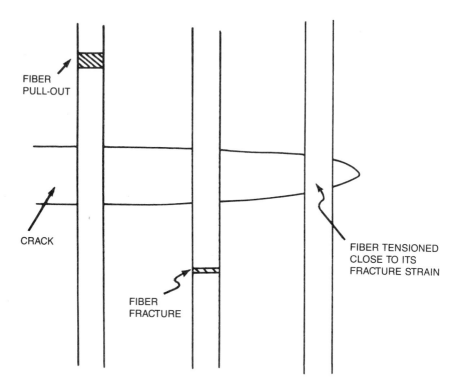

Figure 8.10. Fibers bridging a crack.

fiber located there: typical strains to failure are 0.03 for fiberglass and 0.01 for carbon fiber. Moving away from the crack edge, the displacement gets bigger and fiber fracture and then fiber pull-out are required in order to accommodate the overall crack displacement.

In Equation (8.35), $\gamma_{plasticity}$ may be the dominant term, in which case it is possible for $\mathscr{G}$ to be larger for fracture on a low energy surface or interface. This is the case for basal fracture in sapphire for example. In composites, the fiber/matrix interface can be a low energy interface and yet, at the same time, can give rise to a large contribution to $\mathscr{G}$. The origin of this contribution to $\mathscr{G}$ is interfacial shear strength and friction.

Resistance to fiber/resin interfacial failure may be estimated as follows. Consider in Figure 8.11 a single fiber along the axis of a cylinder of matrix material. Let the composite be tensioned in the fiber direction. At any distance x along the axis, the rate, with respect to x, of transfer of axial load P from matrix to fiber will depend on the difference between the axial displacement u and the axial displacement v that would have existed had the fiber not been there. Following

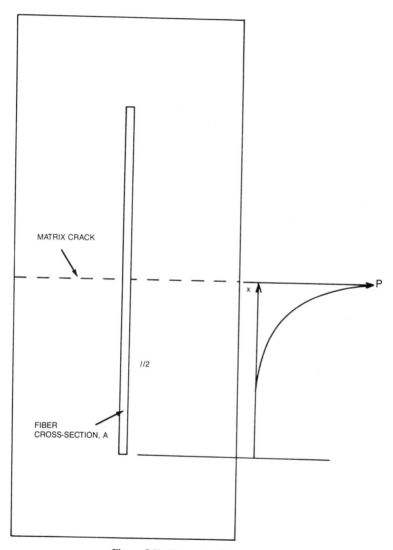

MATRIX CRACK

FIBER
CROSS-SECTION, A

$l/2$

x

P

Figure 8.11. Illustrating fiber pull-out.

H. L. Cox,[31] assume a linear relationship

$$\frac{dP}{dx} = H(u - v) \qquad (8.38)$$

where H is a constant

$$\frac{d^2P}{dx^2} = H\left(\frac{du}{dx} - \frac{dv}{dx}\right) \qquad (8.39)$$

and, since

$$\frac{P}{A} = E\left(\frac{du}{dx}\right)$$

where E is the axial Young's modulus for the fiber, and A is its area of cross-section, Equation (8.39) becomes

$$\frac{d^2P}{dx^2} = H\left(\frac{P}{EA} - e\right) \qquad (8.40)$$

where $e = dv/dx$ is the axial strain in the composite.

Or, writing $P/A' = E'e$, where E' is the axial Young's modulus for the composite and A' is the overall area of cross-section of the composite cylinder,

$$\frac{d^2P}{dx^2} = aP = 0 \qquad (8.41)$$

where

$$a = H\left(\frac{1}{EA} - \frac{1}{E'A'}\right)$$

A simple solution to Equation (8.41) is

$$P = R \sinh\sqrt{a}\,x + S \cosh\sqrt{a}\,x \qquad (8.42)$$

where R and S are constants.

To see the form of the second differential and hence confirm that Equation

[31]H. L. Cox. "The Elasticity and Strength of Paper and Other Fibrous Materials," *Brit. J. Appl. Phys.*, 3:72–79 (1952).

(8.42) satisfies Equation (8.41), consider the function

$$z = \sinh x + \cosh x$$

To find $\dfrac{d}{dx}(\sinh x)$, put

$$y = \sinh x$$

$$= \frac{e^x - e^{-x}}{2}$$

$$\frac{dy}{dx} = \frac{1}{2}\frac{d}{dx}e^x - \frac{1}{2}\frac{d}{dx}e^{-x}$$

$$= \frac{1}{2}e^x - \frac{1}{2}e^{-x}(-1)$$

$$= \frac{e^x + e^{-x}}{2}$$

$$= \cosh x$$

To find $\dfrac{d}{dx}(\cosh x)$, put

$$y = \cosh x$$

$$= \frac{e^x + e^{-x}}{2}$$

$$\frac{dy}{dx} = \frac{1}{2}e^x - \frac{1}{2}e^{-x}$$

$$= \frac{e^x - e^{-x}}{2}$$

$$= \sinh x$$

It is evident that

$$\frac{d^2z}{dx^2} = \sinh x + \cosh x$$

$$= z$$

In the context of fiber pull-out, suppose the matrix has failed at $x = l/2$, see Figure 8.11. The boundary conditions for that half of the fiber in the lower half matrix are $P = 0$ at $x = 0$ and $P = P_{pull\text{-}out}$ at $x = l/2$. $P_{pull\text{-}out}$ is the fiber pull-out load. When $x = 0$, $\sinh \sqrt{a}x = 0$ and $\cosh \sqrt{a}x = 1$, so substitution together with $P = 0$ into Equation (8.42) gives $S = 0$.

When $x = l/2$, Equation (8.42) yields

$$P_f = R \sinh (\sqrt{a})l/2$$

Hence,

$$R = \frac{P_f}{\sinh (\sqrt{a})l/2}$$

and

$$P = P_f \frac{\sinh (\sqrt{a})x}{\sinh (\sqrt{a})l/2} \qquad (8.43)$$

The function $y = \sinh x$ is sketched in Figure 8.12. At small x, $\sinh x \cong x$. At large x, $\sinh x \cong e^x$.

$x < l/2$, so if we take the denominator in Equation (8.43) to be always approximately equal to $e^{\sqrt{a} \cdot l/2}$ and let the numerator approximate to $\sqrt{a} \cdot x$ at small x and to $e^{\sqrt{a} \cdot x}$ at large x, it is evident that the distribution of axial load P will be as shown in Figure 8.11; most of the axial load is carried by that part of the fiber adjacent to its point of emergence from the matrix fracture surface. It is for this reason that pull-out load is effectively independent of fiber length. As J. E. Gordon[32] points out, it is no more difficult for a thrush to pull from the lawn a long worm than it is to extract a short worm.

To give some insight into toughening mechanisms for ceramic matrix composites, it is instructive to examine the origins of toughness of natural composites. Cellulose, the fibrous constituent of wood, is chemically a sugar, and all sugars are mechanically very brittle. Bones and tooth enamel are likewise mainly comprised of mechanically brittle fibers, in fact elongated crystals of calcium hydroxyapatite ($Ca_{10} (PO_4)_6 (OH)_2$), and yet are relatively tough composite materials. Microstructurally, only between 1/10 and 1/3 of the volume of each of these naturally occurring composites is solid material—the rest is open space. It is this open space, or, more precisely, the free surface contributed by this open space that is responsible for the observed high toughness. The point here is that a crack has to stop when it reaches the free surface, so interfacial gaps between fibers in naturally occurring fiber composites certainly are a source of toughness, and mechanically weak interfaces in man-made fiber composites can be also.

---

[32]J. E. Gordon. *Structures*, Penguin, p. 138 (1978).

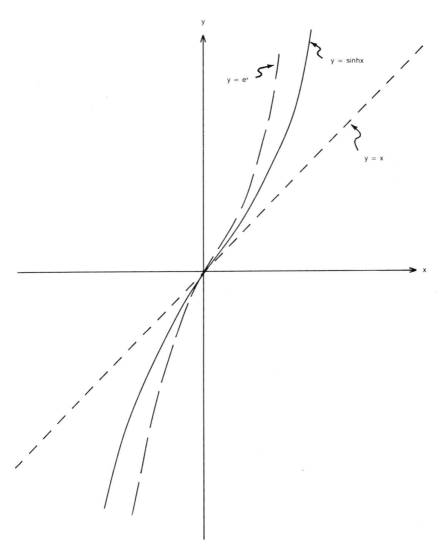

Figure 8.12. Graphical representation (not to scale) of the functions $y = \sinh x$, $y = x$, $y = e^x$.

This crack-stopping property of weak interfaces is exploited in ceramic matrix composites reinforced with mechanically strong fibers, for example, silicon carbide reinforced alumina. Note, however, that exactly the opposite requirement, that is, strong interfacial bonding, is needed in order to realize the toughening action of plastically deformable metal fibers in ceramics. To achieve strong bonding, it is often beneficial to oxidize the metal fiber surface.

8.5 Fracture of a plate subjected to various kinds of edge traction

Inglis[18] considered the case of an elliptic hole in a plate subject to a tensile stress σ applied in a direction making an angle ϕ with the major axis of the ellipse [Figure 8.13(a)] and found that the tensile stress at the surface of and tangential to the hole is

$$\sigma_{\beta\beta} = \sigma \left\{ \frac{\sinh 2\alpha_0 + \cos 2\phi - e^{2\alpha_0}\cos 2(\phi - \beta)}{\cosh 2\alpha_0 - \cos 2\beta} \right\} \qquad (8.44)$$

Griffith[33] uses this solution to explore the tensile stress at the surface of and tangential, that is parallel, to an elliptic cavity in a plate subjected to principal stresses σ_1 and σ_2, respectively making angles ϕ and $\pi/2 - \phi$ with the major axis of the ellipse.

Note that, in his second paper, Griffith has switched from an energy criterion to a stress criterion. The two are differentially related as demonstrated by Frank and Lawn.[34] The strain energy released when a crack grows can be calculated from the imaginary forces needed to close the crack, that is we mend the fracture with, ideally, an infinite number of stitches or rivets pulling the cut edges back into their original positions. We suppose that the material behaves elastically during fracture and during repair so that the stresses are finally what they were before. (To ensure that this assumption is correct in a real experiment, we would have to fracture and repair in small stages so as to avoid things like plastic yielding and the running out of control of the crack). Since the forces exerted by the stitches or rivets are applied normal to the crack faces they are, by definition of a free surface, zero at the start. Hence the stress versus displacement relationship for crack-closing will be as sketched in Figure 8.14(b) and is the mirror image of the same relationship for crack-opening, Figure 8.14(a).

The energy required to close the crack is half of the work done by our

[33]A. A. Griffith. "The Theory of Rupture," *1st Intl. Conf. Appl. Mech.*, Delft, pp. 55–63 (1924).
[34]F. C. Frank and B. R. Lawn. "On the Theory of Hertzian Fracture," *Proc. Roy. Soc.*, A 229:291–306 (1967).

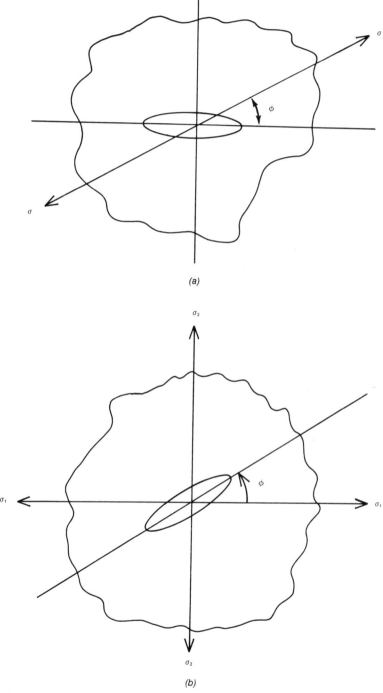

(a)

(b)

Figure 8.13.

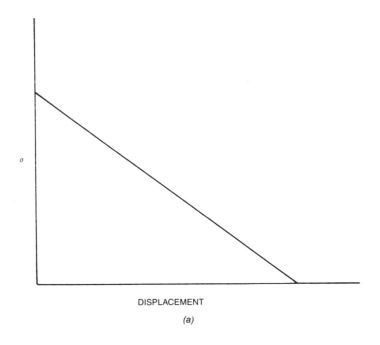

DISPLACEMENT

(a)

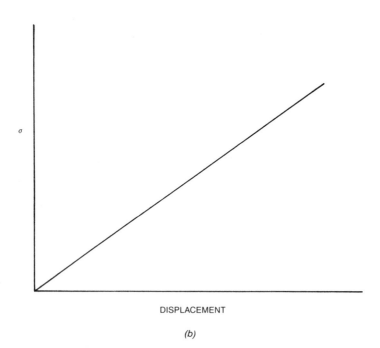

DISPLACEMENT

(b)

Figure 8.14. Stress versus displacement relationships for (a) the growth, and (b) the mending, of a Griffith crack.

imaginary forces, that is

$$dU = \frac{1}{2} \sigma dA \tag{8.45}$$

Therefore

$$\sigma = 2 \frac{dU}{dA} \tag{8.46}$$

Equation (8.46) is the formula for changing from the language of strain energy release to the language of stress.

Referring to Figure 8.13(b), for $\sigma_1 = 0$ the Inglis solution [Equation (8.44)], with the sense of increasing β reversed gives

$$\sigma_{\beta\beta} = \sigma_2 \left\{ \frac{\sinh 2\alpha_0 + \cos 2\phi - e^{2\alpha_0}\cos 2(\phi - \beta)}{\cosh 2\alpha_0 - \cos 2\beta} \right\} \tag{8.47}$$

and for $\sigma_2 = 0$ it gives

$$\sigma_{\beta\beta} = \sigma_1 \left\{ \frac{\sinh 2\alpha_0 + \cos 2(\frac{\pi}{2} - \phi) - e^{2\alpha_0} \cos 2(\frac{\pi}{2} - \phi + \beta)}{\cosh 2\alpha_0 - \cos 2(-\beta)} \right\} \tag{8.48}$$

Hence by superposition,

$$\sigma_{\beta\beta} = \frac{(\sigma_1 + \sigma_2) \sinh 2\alpha_0 + (\sigma_1 - \sigma_2)\{e^{2\alpha_0} \cos 2(\phi - \beta) - \cos 2\phi\}}{\cosh 2\alpha_0 - \cos 2\beta} \tag{8.49}$$

The values of ϕ and β for which $\sigma_{\beta\beta}$ is a maximum are found by differentiation.

Putting $\partial\sigma_{\beta\beta}/\partial\beta = 0$ and taking as solution $\sin 2\beta = A\alpha_0 + 0(\alpha_0^2)$, it can be demonstrated that $\sigma_{\beta\beta}$ is a maximum at two pairs of points on each crack. If $\phi = 0$ or $\pi/2$, these points are at the ends of the major and minor axes respectively. For all other values of ϕ both pairs of points are very near the ends of the axes as sketched for the tensile solution in Figure 8.15. The two extremal values of $\sigma_{\beta\beta}$ always have opposite sign.

To find the extremal values with respect to ϕ, it is necessary to evaluate

$$\frac{\partial \sigma_{\beta\beta}^{extremal}}{\partial \phi} = 0$$

This leads to the following conditions for fracture.

Taking tensile stress as positive and $\sigma_2 > \sigma_1$,

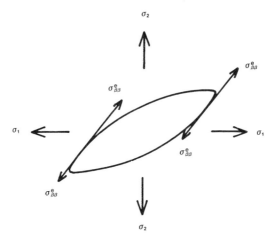

Figure 8.15.

1. If $3\sigma_2 + \sigma_1 > 0$ fracture occurs when $\sigma_2 > K$ where K is the strength in uniaxial tension

   $\phi = 0$, i.e., the fracture surface is perpendicular to σ_2.

2. If $3\sigma_2 + \sigma_1 < 0$ fracture occurs when

$$(\sigma_2 - \sigma_1)^2 + 8K(\sigma_2 + \sigma_1) = 0$$

and

$$\cos 2\phi = -\frac{1}{2} \cdot \frac{(\sigma_2 - \sigma_1)}{(\sigma_2 + \sigma_1)}$$

(8.50)

(8.51)

i.e., crack initiation from the surface of the elliptic cavity occurs at a point near but not at either end of the major axis and is on a plane inclined to the directions of principal stresses, as shown schematically in Figure 8.16.

So Griffith's second paper says that, if it has any shear acting on it, a crack changes its direction, in which case we need a fracture criterion which encompasses the supposition that a crack can change its plane. This is not catered for by the conventional fracture mechanics divisions known as modes I, II, and III, see Figure 8.9.

It is a different mode of elastic stress release which does not open up a crack. In a perfect crystal, for example, fracture modes II and III correspond to the introduction of edge and screw dislocations respectively. In glass, modes II and III leave bonds so stretched that some atoms have different neighbors (such atoms are referred to as frustrated atoms) whereas, in both crystals and glass, fracture mode I leaves some atoms with neighbors on only ore or other side of the crack.

Thus, the specific fracture surface energy for modes II and III is different from that for mode I. If we call the former γ', we expect $\gamma' < \gamma$, where γ is the specific fracture surface energy for mode I. In a crystal, since modes II and III fracture can be modelled by the introduction of dislocations and since planar arrays of dislocations constitute grain boundaries, γ' corresponds to the energy required to introduce a grain boundary—of the order of 0.1–1 kJ m⁻² for most solids. On the other hand, mode I fracture creates free surface so γ more nearly corresponds to the specific surface free energy, which is typically 1–10 kJ m⁻² for most solids. Hence the values of specific fracture surface energy required for a crack to propagate with a continuous decrease of overall energy of the system "specimen plus external forces" is expected to be less for fracture modes II and III than it is for fracture mode I. That is, it is expected that $G_{II} \cong G_{III} < G_I$. In all cases, it is anticipated that the specific fracture surface energy will be lower if time is available for mechanical relaxation of displaced atoms and for water, or other environmental species to dissociate and electrostatically satisfy dangling bonds. Fracture under such conditions is known as sub-critical crack growth (see Section 8.9, question 3).

The two conditions for fracture in two dimensional stress fields are represented graphically[30] in Figure 8.17. If σ_1 and σ_2 are plotted as rectangular coordinates, the second equation is that of a parabola which is concave towards the bisector of negative σ_1 and σ_2. The first equation is that of a vertical tangent to this parabola. When all values of σ_1 and σ_2 are considered, the resulting fracture locus is found to be that shown by the shaded line in Figure 8.17. It is assumed that the material contains Griffith flaws of all orientations and that it fails on that which is most dangerously oriented. Fracture occurs when the point representing the state of stress crosses the locus from its unshaded to its shaded side.

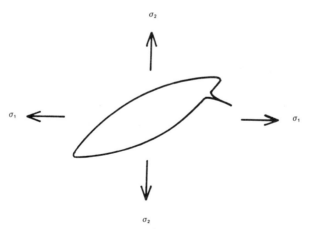

Figure 8.16. Crack propagation from the surface of an elliptic hole (pre-existing crack) in a plate subjected to biaxial stress.

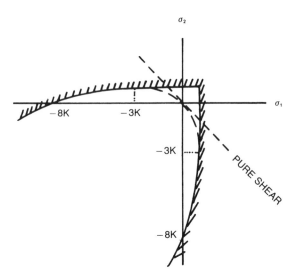

Figure 8.17. Failure envelope for the propagation of a Griffith crack in two-dimensional stress fields. Note that the envelope is open in the compression-compression quadrant.

Fracture envelopes in the form of closed loops are widely used to describe the anisotropy of in-plane strength of laminates. There is no physical justification for this practice; two-dimensional fracture envelopes must be open in the compression-compression quadrant in the manner of Griffith. Layered solids, including laminates, fail during in-plane compression by buckling following delamination between the layers (see Chapter 9).

8.6 Hertzian fracture

So far, we have examined the propagation of cracks in elastic fields which are either homogeneous or weakly inhomogeneous. The best studied example of fracture in a strongly inhomogeneous stress field is that caused by pressing a spherical indenter onto the flat surface of a block of brittle, elastically isotropic material. The nature of the stress field produced by such loading was first analyzed by H. Hertz,[35] and the cone crack which propagates when the normal force P is increased to a critical value P_c, is known as a Hertzian crack. A photograph of a Hertzian crack is shown in Figure 8.18. When the experiment is repeated using spheres of the same material but different radii r, it is observed that, over a large range of radii, P_c is proportional to r (see Figure 8.19). This

[35]H. Hertz. "Über die Berührung fester elastischer Körper," *J. Reine Angew. Math*, 92:156–171 (1881). Reprinted in English in *Hertz's Miscellaneous Papers*, Chapter 5. Macmillan, London.

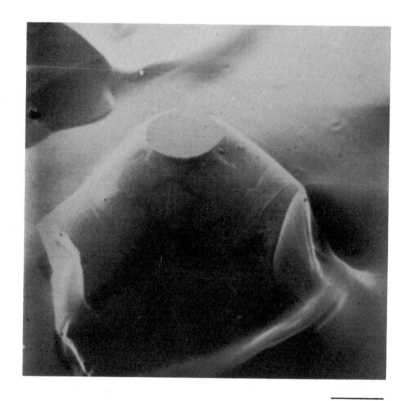

0.5 mm

Figure 8.18. Cone extracted from a Hertzian crack in quartz (after J. W. Heavens).

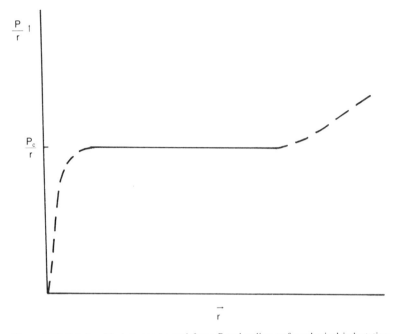

Figure 8.19. Relationship between normal force P and radius r of a spherical indentation.

relationship is known as Auerbach's law[36] and the constant of proportionality (P_c/r) is called the Auerbach constant. This experimental fact that the load upheld by the block at the point of fracture varies linearly with the radius of the indenter suggests that Hertzian fracture is not governed by a stress criterion; a stress criterion would require P to be proportional to r^2.

The elastic field in the specimen is illustrated in Figure 8.20. There is a drop-shaped region below the area of contact in which all three principal stresses are compressive. Beyond this region and just below the surface, Figure 8.20(a), the principal stresses are:

- a tensile radial stress, σ_1
- a compressive hoop stress, σ_2
- a compressive axial stress, σ_3 ($\sigma_3 = 0$ on the surface)

Further below the surface, the directions of σ_1, and σ_3 change as shown in Figure 8.20(b).

In general, $\sigma_1 > \sigma_2 > \sigma_3$. The maximum tensile stress acts across the circle

---

[36]F. Auerbach. "Absolute Hartemessung," *Annalen der Physik und Chemie*, Ser. 3, 43:61–100 (1891).

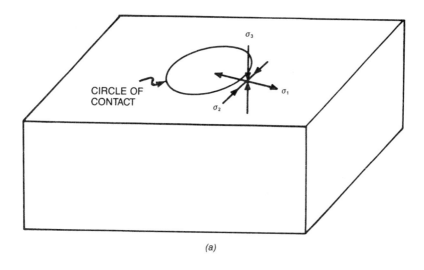

(a)

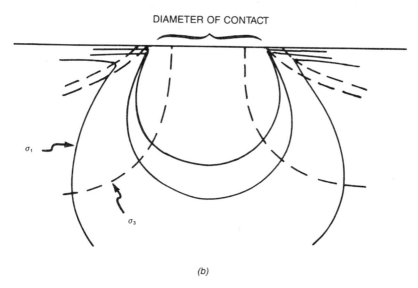

(b)

Figure 8.20. (a) Hertzian stress field. Directions of principal stress just outside circle of contact. (b) Full lines are contours of constant σ_1. Broken lines indicate the direction of σ_3.

of contact and has the value

$$\sigma_1^{max} = (1 - 2\nu) \frac{P}{2\pi a^2} \tag{8.52}$$

where a, the radius of the cricle of contact, is given by

$$a^3 = \frac{4}{3} \cdot \frac{kPr}{E} \tag{8.53}$$

$$k = \frac{9}{16} \left\{ (1 - \nu^2) + (1 - \nu'^2) \frac{E}{E'} \right\}$$

E and E' respectively are the Young's moduli, and ν and ν' respectively, the Poisson's ratios of the specimen and indenter materials. The fact that P_c is proportional to r, and the observation that the Hertzian crack often does not propagate from the circle of contact evidently require investigation.

Comparing Figures 8.18 and 8.20, it is seen that σ_3 stress trajectories leading from near the circle of contact closely resemble the fully developed cone crack, that is the crack is everywhere normal to σ_1, the greatest tensile stress in the elastic field present before fracture. This somewhat surprising fact is the basis of the Hertzian fracture theory proposed by F. C. Frank and B. R. Lawn.[34] The varia-

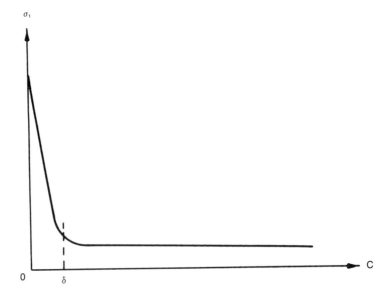

Figure 8.21. σ_1 as a function of crack length c.

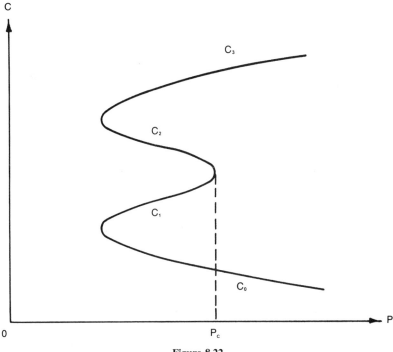

Figure 8.22.

tion of σ_1 along the crack path, Figure 8.21, contains two approximately linear regions and suggests the following four different crack lengths:

(a) a pre-existing sharp surface crack, with length $c << \delta$ experiences a uniform tensile stress. The critical Griffith value, c_0, diminishes as P is increased and, when it becomes less than the length of the pre-existing crack, the latter propagates but not without limit because of the steep decrease in σ_1.

(b) the specimen now contains a stable crack of length $c_1 \cong \delta$, which, under the action of σ_2, runs around the axis of symmetry to form a shallow ring-crack.

(c) a further small increase of c puts the crack into the lower region of uniform stress where it has a second unstable critical length c_2, the magnitude of which is smaller the higher the load.

(d) since the lower region of uniform stress is nearly constant, growth beyond length c_2 leads to further growth to a substantially greater stable length c_3.

The crack length as a function of the normal load is shown in Figure 8.22. Increasing the load on the indenter causes c_1 and c_3 to increase, and c_0 and c_2 to decrease, i.e., c_1 and c_2 approach each other. At the critical load where c_1 and c_2 merge, the crack becomes unstable and spontaneously grows to length c_3. Using a fracture mechanics calculation, it can be demonstrated that the critical event occurs when P/r takes a constant value (Auerbach's constant).

8.7 Impact damage

The laws governing collisions between elastic bodies were first investigated by Newton. The colliding bodies were balls suspended on strings as shown in Figure 8.23. After falling from given heights, the balls struck against each other at the lowest point, and, after rebounding, again reached a certain height. By measuring these heights, and allowing for the resistance of the air, Newton determined the velocities of the balls before and after collision. In this way he showed that when the collision was direct, that is when the relative velocities of the two bodies at the instant of collision was along the common normal at the point of impact, the relative velocity after impact bears a constant ratio to the relative velocity before impact, the relative velocity being of course reversed in direction. Thus, if u, v are the velocities of the bodies before impact, u being the velocity of the more slowly moving body, and U, V are the velocities after impact, then

$$U - V = e(v - u) \qquad (8.54)$$

where e is called the coefficient of restitution. In his experiments Newton found that e depends only on the materials of which the balls are made; it is independent of the masses and relative velocities.

When the initial relative velocity is very large, e is smaller than it is with moderate velocity. In the case of metals which yield, this is attributed to work hardening; in the case of polymers which yield, it is attributed to molecular orientation.

The Hertz calculation tells us that all three principal stresses in the drop-shaped region beneath the indenter are compressive but not equal. Hence there is a maximum shear stress at a distance $\cong a/2$ below the indented surface, where a is the radius of the area of contact. In Figure 8.24, the point of maximum shear

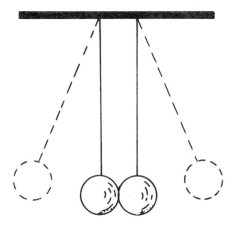

Figure 8.23. Newton's coefficient of restitution experiment.

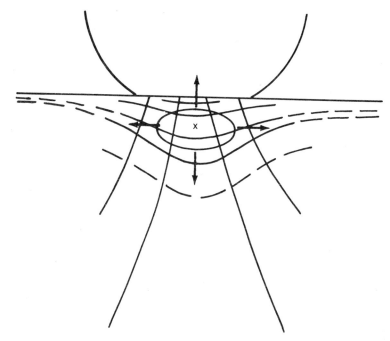

Figure 8.24. Modification of the Hertzian stress field by yield beneath the indentor. Continuous lines denote contours of constant compressive stress, broken lines contours of constant tensile stress.

stress is denoted X. Plastic flow in the region of X would cause a sphere to deform into an oblate spheroid, a vertical section through which is indicated in Figure 8.24 by an ellipse. If the horizontal tensile force is large enough it will more than cancel the horizontal compression due to the Hertzian load and may be large enough to grow radial cracks. From this observation, it is concluded that radial cracks are evidence for plastic deformation during impact.

For some materials densification, rather than yield at constant volume, takes place in the region beneath the indenter. Densification produces the opposite effect, that is enhancement and not cancellation of compression due to the Hertzian load. Reversal of densification on unloading generates tensile stresses which can be large enough to generate unloading cracks.

In fiber reinforced laminates, the lowering of e associated with higher velocity impacts is due partly to delamination—that is to shear cracks in the matrix between plies, see Figure 8.25—and partly, in cases where the laminate bends to large radii, to tensile cracks on the far surface, that is the surface remote from the area of contact by the indenter. At higher velocities still, for example ballistic velocities, impact may result in formation of a crater inside which the impacting body may become lodged.

Returning to Newton's experiment, the momentum of the two bodies is not

Figure 8.25. Internal delamination damage. The impacted surface may reveal no sign of the damage. The far surface may show signs of damage but is often inaccessible.

altered by the impact. So, if m and M are the masses of the two bodies, then

$$mu + Mv = mU + MV \qquad (8.55)$$

Solving Equations (8.54) and (8.55) we find

$$U = \frac{mu + Mv}{m + M} + e \frac{M}{m + M} (v - u)$$

$$V = \frac{mu + Mv}{m + M} - e \frac{m}{m + M} (v - u)$$

Hence,

$$\frac{1}{2} mU^2 + \frac{1}{2} MV^2 = \frac{1}{2} mu^2 + \frac{1}{2} Mv^2 - \frac{1}{2} (1 - e^2) \frac{Mm}{M + m} (v - u)^2 \qquad (8.56)$$

from which it is evident that the kinetic energy after impact is less than the kinetic energy before impact by

$$\frac{1}{2} (1 - e^2) \frac{Mm}{M + m} (v - u)^2 \qquad (8.57)$$

If e is unity, there is no loss of kinetic energy. In all other cases there is a finite loss of kinetic energy, some of it being transformed during the collision into heat; a small part of it may, in some cases, be spent in throwing the balls into vibration about their figures of equilibrium; and, for higher velocity impacts, much of the kinetic energy lost is consumed by the processes of work hardening and fracture.

For a particle of small mass colliding with a stationary target, Equation (8.57) reduces to

$$1/2 \ (1 - e^2) \ mv^2 \qquad (8.58)$$

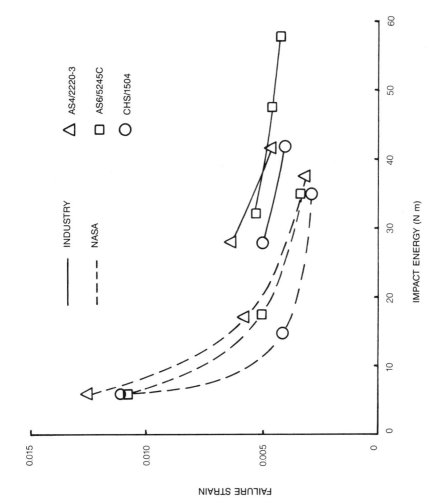

Figure 8.26. In-plane compressive strain to failure after impact (after J. C. Halpin).

Small localized delamination within a laminated plate has little effect on in-plane tensile strength, but seriously impairs the in-plane compression strength. The data reproduced in Figure 8.26 show the latter effect, measured by compressive strain to failure for carbon fiber reinforced epoxy resin laminates. To understand this observation, consider tensile and compressive deformation of a laminated plate. One small region of weak interlaminar bonding has no effect on in-plane tensile strength because, at the weakness, we effectively have two plates and the tensile strength of two (identical) plates is twice the tensile strength of one plate. On the other hand, in-plane compression of two plates is subject to buckling phenomena. Even when tested by in-plane compression between external plates introduced to prevent macroscopic buckling (refer to ASTM D-695), the region of weakness can still delaminate and buckle on the microscopic scale; the plates may not be able to bulge towards each other, but they will be able to bulge apart. In accordance with Euler's theory for buckling columns, this local buckling will be proportional to some power of the plate thickness.

One practical way of constraining buckling not to occur, including out-of-plane displacements associated with free-edge effects (see Chapter 9, Section 9.8), is through-thickness stitching. Examples of stitching patterns used on carbon fiber reinforced laminates are shown in Figure 8.27. Tests demonstrate that the higher the stitching density, the smaller the overall area of delamination due to impact.

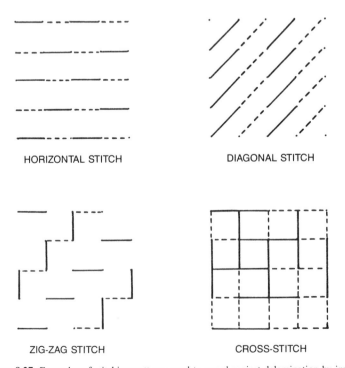

HORIZONTAL STITCH   DIAGONAL STITCH

ZIG-ZAG STITCH   CROSS-STITCH

Figure 8.27. Examples of stitching patterns used to guard against delamination by impact.

Finally, the energy absorption characteristics of honeycomb cores, particularly aluminum honeycomb, should be noted. Microstructural damage by modest impact is rarely seen on the back plate of a laminate plate sandwich core structure. Nor is there usually any damage to the adhesive bond between the core and backplate.

8.8 Erosion and ablation

Erosion is an important branch of impact damage. Erosion is initiated by cavitation, see Figure 8.28. For each material and eroding medium combination, there exists a minimum intensity of pressure wave above which cavitation occurs. The erosion of materials by water,* including the problem of damage to aerospace materials by impinging raindrops, can be simulated in the laboratory by immersing the material in water and subjecting it to a beam of ultrasound. It is assumed that each cavity is nucleated at an air bubble which then becomes trapped within it. In water, sound propagates by way of compressional waves at a velocity (v) of 1410 m s^{-1}. The wavelength of, for example, 20 kHz (ν) vibrations is therefore

$$\lambda = \frac{v}{\nu} = \frac{1410 \ (\text{m/s})}{20 \times 10^3 \ (1/\text{s})}$$

$$= 7.05 \ \text{cm}$$

Since this value is considerably larger than the thickness of most laminates, it can be assumed that the associated pressure wave is space independent. The period of vibration is $1/(20 \times 10^3)$ seconds (1/20 ms), so the tensile phase lasts for 1/40 ms during which time the bubble grows and emits a compressional wave, and if it could, this wave would travel 7.05 cm before any further bubble nucleation occurs. In a perfect spherically symmetric bubble, the ensuing water jets would be radially opposed. Within a laminate the spherical symmetry is destroyed, resulting in an imbalance between the impinging jets which causes the jets to be growing outward at the end of the implosion. The hammer blow on the base of the cavity at each contraction is conveyed by the air cushion and the pressure is immense (of the order of 1 GN m^{-2}). The air cushions the workpiece against the devastating attack of the high pressure water jets. As the wave front travels down the cavity, it will experience viscous drag by the cavity walls and may change from being concave downward to convex downward so that cushioning is effective only in a ring of air.

Erosion by particles of solid material, including the erosion of re-entry vehicle materials by dust particles and hailstones, has been studied at the microstructural

---

*NB: Fiber prominence, the protrusion of fiber ends on weathered surfaces, is a matrix dissolution phenomenon and has little to do with erosion proper.

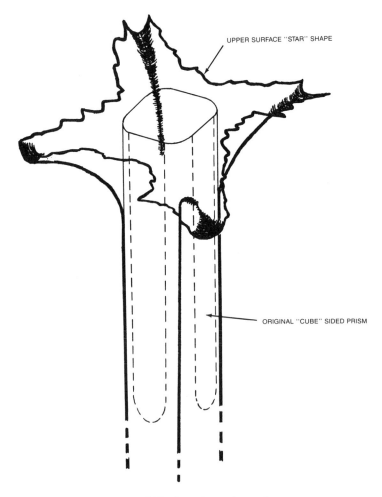

UPPER SURFACE "STAR" SHAPE

ORIGINAL "CUBE" SIDED PRISM

Figure 8.28. Illustrating cavity erosion.

level. For three-dimensional carbon/carbon composites, impact by high velocity particles creates craters, see Figure 8.29. The growth of radial cracks between fibers oriented perpendicular to the free surface is inhibited by the constraint, indicated by arrows in Figure 8.29(a), that is offered by the adjacent material. Instead, the fibers buckle in such a way that the overall volume remains as near constant as possible, a condition met if the fiber bundles crease at about 45° to the bundle. This condition is illustrated in Figure 8.29(b). Formation of the crease, or kink band as it is sometimes called, is matrix dominated; the near 45° mathematical planes are planes of maximum shear stress and whether or not a crease can form is determined by the shear strength of the matrix.

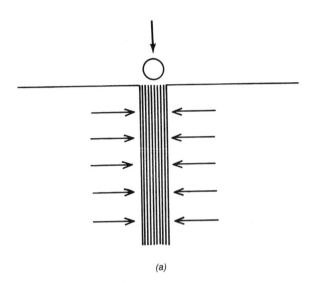

(a)

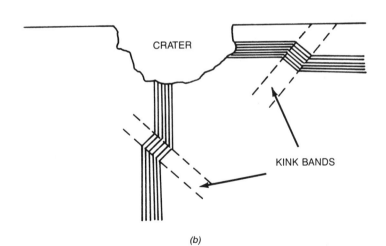

CRATER

KINK BANDS

(b)

Figure 8.29.

Related to solid particle erosion is the phenomenon of ablation, the incidence of which requires the large stress exponents characteristic of a non-quadratic dependence of strain energy density on load. There are several possibilities. Models based on impact by spherical particles rely on the non-constant geometry of an area of contact that increases with load. The mean strain energy density is not important; it is the strain energy density at the surface of contact that gives rise to stress induced melting and burning, and in the Hertzian stress field, it is strongly non-linear. Other particle shapes give different results. Thus, for a wedge shaped particle with its edge pressing against a planar surface, the problem is one of large strain elasticity; the stress must approach infinite magnitude at the edge, since the stress in the wedge is proportional to its thickness.

8.9 Worked examples

1. Use Griffith's criteria for fracture in a plate subjected to biaxial stress, Equations (8.50) and (8.51), to predict failure in (a) torsion, (b) uniaxial compression and (c) pure shear with a superposed hydrostatic pressure.

(a) Torsion, that is pure shear. Remembering the convention that $\sigma_2 > \sigma_1$, the principal stresses are $\sigma_2 = \sigma$ and $\sigma_1 = -\sigma$. Griffith's criterion (i), Equation (8.50), applies and predicts fracture when $\sigma = K$. This condition is identical to fracture in uniaxial tension; that is, the shear strength is equal to the tensile strength. Fracture in torsion is expected to occur on the surface perpendicular to σ_1, that is, on the helical surface inclined at $45°$ to the axis of torsion, see Figure 8.30(a).

(b) Uniaxial compression. $\sigma_2 = 0$, $\sigma_1 = -\sigma$ so criterion (ii), Equation (8.51), applies and fracture is predicted when $\sigma_1 = -8K$; that is, the compressive strength of a brittle material is expected to be eight times its tensile strength and fracture occurs on the surface for which $\cos 2\phi = (-1/2)$, that is at $60°$ to the axis of compression. Note that if the compression platen surfaces are not lubricated, barrelling is likely and may initiate longitudinal cracking; that is, fracture may in practice occur parallel to the compression axis, see Figure 8.30(b).

  If the compressed solid is a uniaxial composite oriented with fiber direction along the axis of compression, then unless the fibers are close packed, the matrix is unlikely to be sufficiently rigid to prevent fiber buckling. This phenomenon is an example of classical Euler buckling of axially compressed columns, for which the wavelength of buckling is proportional to the fiber diameter. The fibers may buckle in-phase or out-of-phase as sketched in Figure 8.30(c). In both cases the failure is matrix dominated, by the shear strength of the matrix material for the in-phase case, and by the tensile strength of the matrix material in the out-of-phase case. The upshot is that the uniaxial compressive strength of uniaxial composites is less than the uniaxial tensile strength.

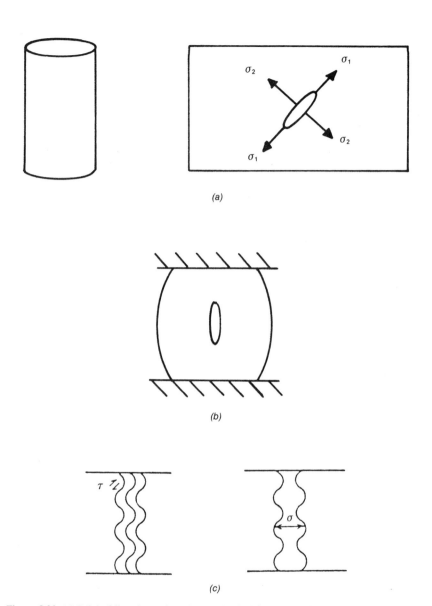

Figure 8.30. (a) Brittle failure in torsion. (b) Longitudinal failure in uniaxial compression between unlubricated platens. (c) Fiber buckling modes in uniaxial composite subjected to uniaxial compression parallel to the fiber direction.

$$\sigma_{compression} = \frac{G_{matrix}}{1 - \eta} \qquad \text{in-phase} \qquad (8.59)$$

$$\sigma_{compression} = \frac{2}{3}\,\eta\,\frac{\eta\,E_m E_f}{1 - \eta} \qquad \text{out-of-phase} \qquad (8.60)$$

where η is the fiber volume fraction, where G_{matrix} is the shear modulus of the matrix material, and E_m and E_f are the values for Young's modulus of the matrix and fiber materials respectively.

There is the following important lesson to be learned from nature about guarding against compression failure of uniaxial composites. A column fabricated as a uniaxial composite, resin bonded fiberglass for example, behaves as a Hookean solid. Unloaded, it is stress-free as indicated by the horizontal line in Figure 8.31(a). If it buckles under load, in the direction of the arrows in Figure 8.31(a) then a distribution of axial stress, changing from compression on one side to tension on the opposite side, is created and fiber buckling in the compressed region is a likely failure mode.

To guard against this failure, trees are heavily pre-stressed in the manner indicated by the broken line in Figure 8.31(b); the outer layers are tensioned by as much as 10–13 MN m^{-2}. On bending in the wind, the distribution of axial stress in a tree changes to that sketched in Figure 8.31(c); the largest compressive stress is relatively small and compression failure is usually avoided.

(c) Pure shear with a superposed hydrostatic pressure. The condition for fracture under these conditions is of special importance in the design of submarine hulls.

$$\sigma_2 = \sigma - \sigma', \quad \sigma_1 = -\sigma - \sigma'$$

where $-\sigma'$ is the hydrostatic pressure.

If $\sigma' > \sigma/2$, $\sigma_2 < \sigma/2$ and $\sigma_1 > -\frac{3}{2}\sigma$. Criterion (ii), Equation (8.51), applies and fracture occurs if

$$(<\sigma' + >3\sigma')^2 + 8K(<\sigma' - >3\sigma') = 0$$

i.e., if

$$(<4\sigma_1)^2 - 8K(-<2\sigma') = 0$$

or

$$\sigma' > K$$

For most solids, since K is of the order of 1 kilobar, a superposed pressure of $\cong 1$ kbar would prevent all possibility of failure by shear.

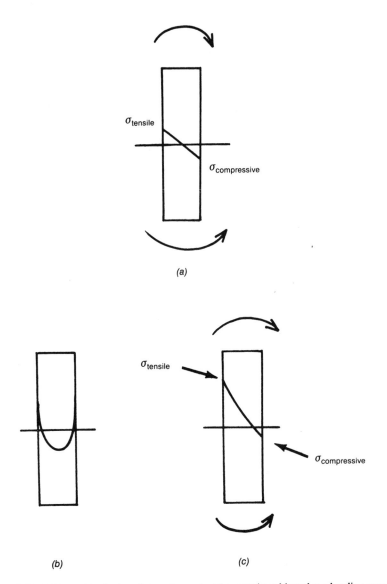

Figure 8.31. (a) Linear distribution of stress in a uniaxial composite subjected to a bending moment. The maximum tensile and compressive stresses are equal in magnitude. (b) Pre-stressing in a tree. The outer wood is axially tensioned, and the inner wood is axially compressed. (c) Pre-stressed tree subjected to a bending moment. The stress distribution is non-linear, and the maximum compressive stress is substantially smaller than it was in (a).

2. Most naturally occurring cracks are opening mode cracks, that is they grow perpendicular to the line of action of tensile stress. If the state of stress in a material containing such a crack changes so that shear stress now acts on the propagating crack, the crack changes its plane of propagation (see Section 8.5). Not only that, but the crack is equally likely to propagate on either of two planes, inclined one above and the other below the original plane of propagation. Explain this observation.

In the case of an internal crack, the two new orientations for crack propagation originate at the two points shown in Figure 8.15.

For a crack propagating from a surface flaw, the existence of two new orientations for the crack can be argued as follows. In accordance with Equation (8.30), the strain energy release rate ($\mathscr{G}$) is proportional to σ^2 for a crack propagating under the action of a tensile stress σ. By the same token, $\mathscr{G}$ for a crack propagating in a stress field arising from application of a shear stress τ is proportional to τ^2. In both cases, the equations for strain energy density are equations of linear elasticity. Hence, for a crack propagating under the combined action of tensile and shear stresses, the strain energy release rate must be proportional to $\sigma^2 + \tau^2$, a quantity which is identically the same for the states of stress (σ, τ) and $(\sigma, -\tau)$. These two states of stress, shown on a Mohr circle construction in Figure 8.32, act across planes inclined to the principal axes by $\pm\theta$.

Thus, there exists a bi-stability in respect to choice of fracture plane for a crack propagating in an elastic field that has finite normal and shear components of stress.

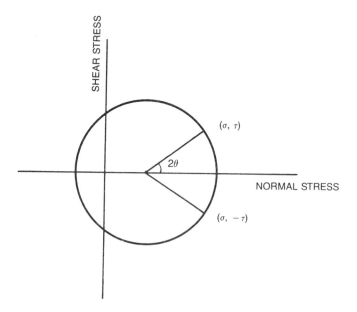

Figure 8.32.

3. The phenomenon of slow crack growth at stresses smaller than the Griffith fracture strength (i.e., at stress intensity factors smaller than $\mathcal{K}_{critical}$) is attributable to environmental effects, usually water adsorption at the crack edge. Derive a mathematical model to characterize this phenomenon, and to predict the time to failure (refer to A. G. Evans[37]).

The linear elastic fracture mechanics stress intensity factor is

$$\mathcal{K} = \sigma(\pi a)^{1/2} \tag{8.61}$$

where σ is the applied stress ($\sigma = \sigma_{11}$ for mode I fracture, $\sigma = \sigma_{12}$ for mode II fracture, and $\sigma = \sigma_{11} + \sigma_{12}$ for mixed modes I plus II fracture, etc), and a is the critical Griffith crack length.

If $d\sigma/dt = 0$, then

$$\frac{d\mathcal{K}}{dt} = \frac{1}{2}\,\sigma\pi^{1/2}a^{-1/2}\,\frac{da}{dt}$$

$$= \frac{\mathcal{K}v}{2a}$$

where the crack velocity

$$v = \frac{da}{dt}$$

is assumed to be constant.

Hence

$$\frac{d\mathcal{K}}{dt} = \frac{\pi\sigma^2}{2}\,\frac{v}{\mathcal{K}}$$

And, the time for crack propagation is

$$t = \frac{2}{\pi\sigma^2}\int_0^\infty \frac{\mathcal{K}}{v}\,d\mathcal{K}$$

This integral is sketched in Figure 8.33.

It is evident that the range of $\mathcal{K}$ values from zero to $\mathcal{K}_0$, indicated by a brace in Figure 8.33, provides an infinite contribution to the integral! This complication is circumvented in practice by evaluating the integral from $\mathcal{K}_0$ to $\mathcal{K}_c$.

---

[37]A. G. Evans. "A Simple Method for Evaluating Slow Crack Growth in Brittle Materials," *Int'l. J. Fracture*, 9:267–275 (1973).

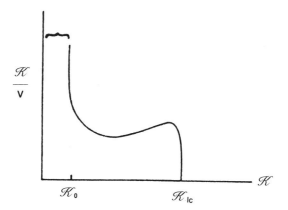

Figure 8.33. Graphical representation of the integral $\int \dfrac{\mathcal{K}}{v}\, d\mathcal{K}.$

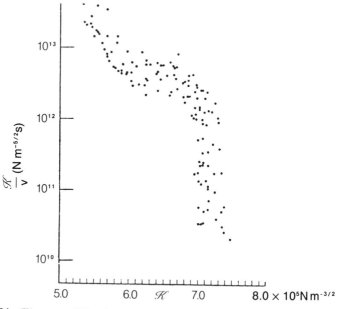

Figure 8.34. $\mathcal{K}/v$ versus $\mathcal{K}$ data for mode I fracture of soda lime glass (after S. M. Wiederhorn).

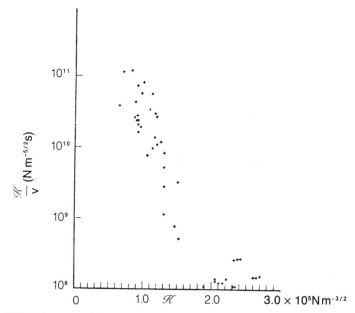

Figure 8.35. $\mathcal{K}/v$ versus $\mathcal{K}$ data for mode III fracture of soda lime glass (after D. A. Tossell).

$\mathcal{K}/v$ versus $\mathcal{K}$ data for mode I and mode III fractures in soda lime glass exposed to identical aqueous environments are shown in Figures 8.34 and 8.35 respectively. Note that, for a given crack velocity, $\mathcal{K}_{III} < \mathcal{K}_{I}$.

8.10 Further examples

1. While trying to hammer a nail into a plank of wood in a confined space such that the swing of the hammer is restricted to about 50 mm it was found that the nail did not move perceptibly. Assuming that the same force or couple could be exerted on the handle of any hammer, is there an optimum mass for the head? Does it matter whether the plank is supported from behind or at its ends (University of Bristol, 1976)?

2. Figure 8.36 shows the fracture surface for a uniaxial glassfiber reinforced resin composite. Assuming that the broken fibers were not loose in their "sockets" show that the contribution to the overall work of fracture offered by fiber pull-out is of the order of $\pi\tau dl_c^2$ where τ is the interfacial shear strength, l_c is the critical fiber length (see Chapter 2, Section 2.8) and d is the fiber diameter.

3. Figure 8.37 shows the fracture surface for another uniaxial glassfiber reinforced resin composite. Here the crack propagated continuously from fiber to

Figure 8.36. Fiber pull-outs and sockets created during tensile fracture of glassfiber reinforced epoxy resin (after A. Kelly).

A

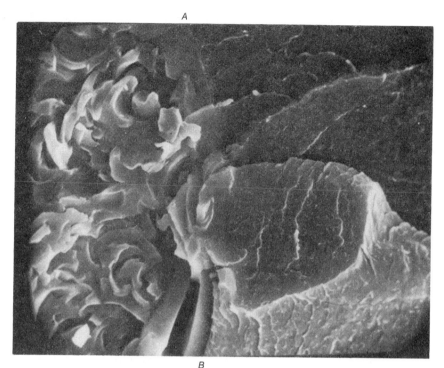

B

Figure 8.37. Brittle fracture surface for a glassfiber reinforced resin (scanning electron micrograph courtesy of Z. R. Xu).

matrix, that is without any fiber pull-out. Across the fiber/resin interface AB the hackle markings are continuous with more or less no change in curvature, from which it is deduced that there was no change in crack velocity. By what proportion must the strain energy release rate ($\mathscr{G}$) have changed at the interface? Take as values for Young's modulus 1 GN m^{-2} for fiberglass and 100 MN m^{-2} for resin. Take as values for Poisson's ratio 0.2 for fiberglass and 0.39 for resin. When the composite failed, there was a loud audible report. Which material (fiber or matrix) contributed more energy to this report?

4. Some thermoplastics are toughened by the introduction of glass fibers (amorphous polyethersulfone is one example), whereas others are embrittled (semi-crystalline polyoxymethylene and semi-crystalline, particle toughened polyethylene terephthalate are examples). Explain.

5. Poor crashworthiness poses a serious limitation to the application of fiber reinforced plastics. List the physical mechanisms which give rise to crashworthy characteristics in more conventional materials, and discuss those mechanisms which might be exploited in the development of metal matrix composites.

6. In a laminate subjected to alternating loads, the protection against failures, due to easing off of the local stresses at the edge of a crack effected by plastic yield, is limited. Explain.

7. What would you look for as evidence of fiber creasing (buckling) in a scanning electron microscope study of the fracture surface of a carbon fiber/epoxy resin composite that has failed in compression? Illustrate your answer with a sketch. Would it be possible to design, say, a uniaxial composite that is not susceptible to fiber creasing when loaded by compression in the fiber direction?

8. Is anything gained or lost by proof-testing a composite material?

9

Anisotropy of strength

9.1 Airy stress functions in the air[38]

Fiber reinforced laminates are widely used in the form of plates, beams and rods, and the easiest way to solve stress problems involving plates, beams and rods is to construct the Airy stress function in the air. The Airy stress function (χ) is a scalar function of position in two dimensions whose second derivatives will represent the components of stress for any two-dimensional or quasi-two-dimensional structure in equilibrium.

$$\left. \begin{array}{cc} \sigma_{xx} = \dfrac{\partial^2 \chi}{\partial y^2} & \sigma_{yy} = \dfrac{\partial^2 \chi}{\partial x^2} \\[3mm] \sigma_{xy} = \sigma_{yx} = -\dfrac{\partial^2 \chi}{\partial x\, \partial y} \end{array} \right\} \tag{9.1}$$

Note that the σ_{ij} are two-dimensional stresses, that is they have the dimensions of force per unit length. It is important to remember that the Airy stress function deals only with stresses in equilibrium, putting aside all consideration of strain. In passing, although this point will not be pursued here, it might be noted that, in the case of homogeneous isotropic elasticity at least, Hooke's law would give the additional biharmonic equations $\nabla^4 \chi = 0$.

9.2 Worked examples

1. Figure 9.1 is a loaded cantilever beam which can be regarded as a truss of pin-jointed rods of zero weight. Construct its Airy stress function in the air, and

[38]F. C. Frank. "Airy Functions in the Air: An Easy Way with Stress Problems," *Phys. Educ.*, 13:258–263 (1978).

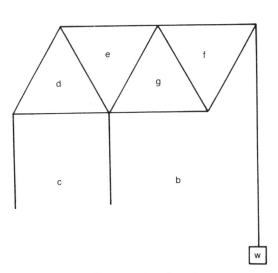

Figure 9.1. Loaded cantilever beam.

calculate the stresses in each member if all rods are of length 1 m and the load supported by the beam is 40 kN.

In the air, there is no stress, that is all second differentials of χ are zero. Hence, if χ is represented by a surface having elevation χ (x,y), that surface must be a plane wherever we have only air present. Since the gradient of this plane is of no significance, we are free to make it zero at one place, in the area labelled a for example. The absolute value of χ likewise has no significance, so we will also make it zero in area a.

In string ab (rods and string will be identified by the areas either side of them) there exists a tension σ_{yy}, and across this string

$$\int_b^a \sigma_{yy} \, dx = W \tag{9.2}$$

where W is the weight supported by the beam. From Equation (9.1),

$$W = \int_b^a \frac{\partial^2 \chi}{\partial x^2} \, dx$$

$$= \left(\frac{\partial \chi}{\partial x}\right)_a - \left(\frac{\partial \chi}{\partial x}\right)_b$$

but, since we have put $(\partial \chi/\partial x)_a = 0$, $(\partial \chi/\partial x)_b = -W$ that is, the slope of χ with respect to x in area b is negative.

Since there is no shear stress across the string, slope $(\partial \chi/\partial y)_b$ is zero, so the χ surface evidently changes gradient by a sharp valley-fold at the string. In Figure 9.2, this change is represented by equispaced straight contour lines of constant χ labelled 0 to 60 kN m, appropriate for rods of length 1 m and a load $W = 40$ kN.

$$\left(\frac{\partial \chi}{\partial x}\right)_b = \frac{-60 \ (\text{kN})}{\frac{3}{2} \ (\text{m})}$$

is the horizontal change in slope in area b.

This brings us to pillar bc, where there must exist a ridge-fold (compressive stress in the pillar) since we have to get down to $\chi = 0$ at pillar ac at which there is a valley-fold (tension).

Because of the pin-joints, the rods experience only axial compression or tension. This means that they must delineate ridge- or valley-folds in the χ surface, and that χ contours must be continuous across them. That makes it very easy to

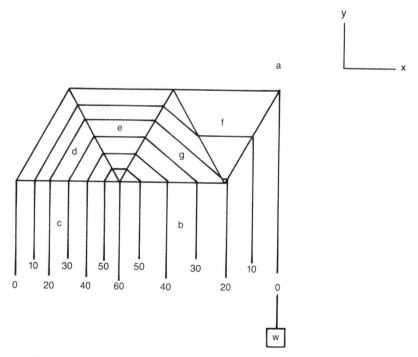

Figure 9.2. Contours of constant χ in the air between the members of Figure 9.1.

construct the χ contours in areas d, e and f. In area g, the contours are found by joining the ends of χ contours belonging to the adjacent areas.

The stresses in any member are found by calculating the change of slope of a line at right angles to that member. There is no need to resolve stresses on to our arbitrarily oriented coordinate axes x,y. Thus the slope at right angles to pillar bc changes from -40 kN m^{-1} in area b to $+60$ kN m^{-1} in area c, so the stress in bc is $(-40 - 60) = -100$ kN m^{-1}. Table 9.1 lists the individual stresses.

2. Modern cross-bows are of fiber reinforced laminate construction. Prove the following facts for a tensioned cross-bow.

(1) The change across the bow of the value of χ is the bending moment at the crossing plane.

Table 9.1.

ab	0 in a	$-\dfrac{60}{(3/2)}$ in b	$0 + 40 = +40$
bc	-40 in b	$+60$ in c	$-40 - 60 = -100$
ac	$+60$ in c	0 in a	$+60 - 0 = +60$
ad	$+\dfrac{60}{(\sqrt{3/2})}$ in d	0 in a	$+\dfrac{120}{\sqrt{3}} - 0 = +\dfrac{120}{\sqrt{3}}$
ae	$+\dfrac{60}{(\sqrt{3/2})}$ in e	0 in a	$+\dfrac{120}{\sqrt{3}} - 0 = +\dfrac{120}{\sqrt{3}}$
af	$+\dfrac{20}{(\sqrt{3/2})}$ in f	0 in a	$+\dfrac{40}{\sqrt{3}} - 0 = +\dfrac{40}{\sqrt{3}}$
bg	0 in b	$+\dfrac{40}{(\sqrt{3/2})}$ in g	$0 - \dfrac{80}{\sqrt{3}} = -\dfrac{80}{\sqrt{3}}$
cd	0 in c	$+\dfrac{30}{(\sqrt{3/2})}$ in d	$0 - \dfrac{60}{\sqrt{3}} = -\dfrac{60}{\sqrt{3}}$
de	$-\dfrac{30}{(\sqrt{3/2})}$ in e	$+\dfrac{30}{(\sqrt{3/2})}$ in d	$-\dfrac{60}{\sqrt{3}} - \dfrac{60}{\sqrt{3}} = -\dfrac{120}{\sqrt{3}}$
eg	$-\dfrac{10}{(\sqrt{3/2})}$ in g	$+\dfrac{30}{(\sqrt{3/2})}$ in e	$-\dfrac{20}{\sqrt{3}} - \dfrac{60}{\sqrt{3}} = -\dfrac{80}{\sqrt{3}}$
gf	$-\dfrac{50}{(\sqrt{3/2})}$ in g	$+\dfrac{10}{(\sqrt{3/2})}$ in f	$-\dfrac{100}{\sqrt{3}} + \dfrac{20}{\sqrt{3}} = -\dfrac{80}{\sqrt{3}}$
bf	$-\dfrac{10}{[1/(2\sqrt{3})]}$ in b	$+\dfrac{10}{(\sqrt{3/2})}$ in f	$-20\sqrt{3} - \dfrac{20}{\sqrt{3}} = -\dfrac{80}{\sqrt{3}}$

(2) The change across the bow of the longitudinal gradient of χ is the shear force it transmits, and also the rate of change along it of bending moment.

Figure 9.3 represents a tensioned cross-bow.

The χ-map has two equal valley-folds (tension) on lengths ab and ac of the string, and these require there to be a ridge-fold (compression) across the stem bc.

Straight line contours of $\chi = 1,2,3$ are drawn at an arbitrarily chosen equal spacing. The scale of χ is determined by the fact that the change of slope $(d\chi/dx)$ on a straight line which crosses the stem at right angles — the thrust in the stem — equals the pull required to tension the bow. On the bow itself, we have a new situation. There exists a change across it of the gradient of χ parallel to it. Also, except at its ends, the magnitude of χ changes across the bow. To see the implications of these facts, consider in Figure 9.4 a detail of Figure 9.3.

The local coordinate axes have been oriented with the y-axis parallel to the χ contours so that $(d\chi/dx) = 0$ in the air (spaces c and a). Note that χ may vary in a complicated way within the fiber reinforced laminate of the bow itself; we know nothing of the details of this variation and have no need of this knowledge for the present purposes. Integration along the y-axis, line OP, gives us the tensile force acting across OP:

$$
\begin{aligned}
T_{OP} &= \int_{O}^{P} \sigma_{xx}\, dy \\[2mm]
&= \int_{O}^{P} \left(\frac{\partial^2 \chi}{\partial y^2} \right) dy \\[2mm]
&= \left[\frac{\partial \chi}{\partial y} \right]_{P} - \left[\frac{\partial \chi}{\partial y} \right]_{O} \qquad\qquad (9.3) \\[2mm]
&= 0
\end{aligned}
$$

The shear force acting across OP is:

$$
\begin{aligned}
S_{OP} &= \int_{O}^{P} \sigma_{xy}\, dy \\[2mm]
&= -\int_{O}^{P} \left(\frac{\partial^2 \chi}{\partial x\, \partial y} \right) dy \\[2mm]
&= \left[\frac{\partial \chi}{\partial x} \right]_{O} - \left[\frac{\partial \chi}{\partial x} \right]_{P} \qquad\qquad (9.4)
\end{aligned}
$$

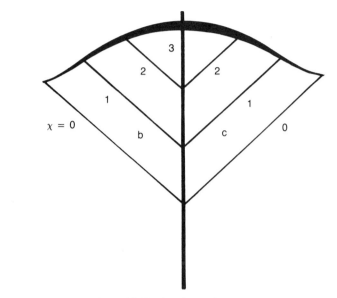

a

Figure 9.3. Tensioned cross-bow.

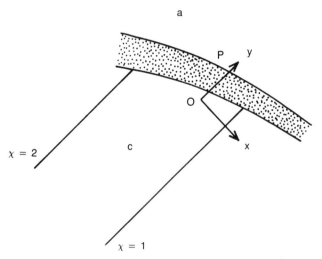

Figure 9.4. Detail from Figure 9.3. Axis Oy is parallel to the χ contours.

It is evident from Figure 9.4 that $(\partial\chi/\partial x)_o$ is negative.

$$\left(\frac{\partial\chi}{\partial x}\right)_P = 0$$

Similarly, by integrating along OP, we obtain the bending moment:

$$M_{OP} = \int_o^P y\,\sigma_{xx}\,dy$$

$$= \int_o^P y\left(\frac{\partial^2\chi}{\partial y^2}\right)dy$$

$$= \left[y\left(\frac{\partial\chi}{\partial y}\right)\right]_o^P - \int_o^P \left(\frac{\partial\chi}{\partial y}\right)dy \qquad (9.5)$$

$$= 0 - [\chi]_o^P$$

$$= \chi_o - \chi_P \qquad (9.6)$$

Had we selected another orientation for OP, we would have obtained different answers. For example, suppose we rotate our coordinate axes counter-clockwise until the y-axis, and our test-line OP, is at right angles to the bow. $(\partial\chi/\partial y)_o$ would become positive, $(\partial\chi/\partial y)_P$ would remain zero, thereby making T_{OP} negative (compression). $-(\partial\chi/\partial x)_o$ becomes smaller and $(\partial\chi/\partial x)_P$ remains zero, whether we rotate OP counter-clockwise or clockwise; our initial choice of orientation of OP in fact found the maximum shear force S_{OP}.

The first term in Equation (9.5) becomes negative (instead of zero) for counter-clockwise rotation (positive for clockwise rotation) of OP, and represents the (clockwise) moment about O of the tensile force exerted across OP upon the material to the left. The second term does not alter with rotation of our coordinate axes; it represents the pure bending moment in the bow at the plane where it is crossed by OP and that place does not move very much if the bow is thin and if origin O is close to it.

3. Fiber reinforced prepregs readily lend themselves to the fabrication of curved laminated beams and panels. The through-the-thickness tensile strength of laminates is typically as low as 5% of the in-plane tensile strengths; 50 MN m^{-2} compared with 2 GN m^{-2}, for example, in the case of high modulus carbon fiber/epoxy resin laminates. Construct the Airy stress function in the air for a uniformly curved beam subjected to a uniform bending moment tending to straighten it, and hence show that, for a beam of thickness 10 mm bent to a mean

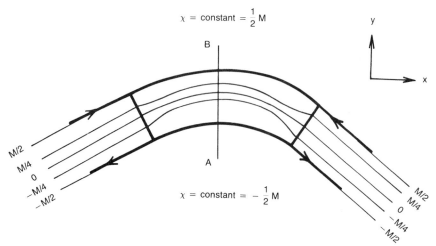

Figure 9.5. χ-map for a uniformly curved beam with a uniform bending moment M tending to straighten it.

radius of 55 mm, the radial stress as a percentage of the maximum longitudinal stress is also of order 5%.

The map of χ contours is sketched in Figure 9.5. χ in the air is constant and equal to $-M/2$ inside the curve, and constant and equal to $+M/2$ outside the curve. Within the material of the beam χ must vary radially to make a transition between these two values, with no discontinuity in χ or its gradients at the inner and outer boundaries, viz $r = r_0 - \frac{1}{2}h$ and $r = r_0 + \frac{1}{2}h$. h is the thickness of the beam.

The simplest function satisfying these conditions is

$$\chi = M\{3(r - r_0)/h - 4(r - r_0)^3/h^3\} \tag{9.7}$$

for which

$$\frac{\partial \chi}{\partial r} = \frac{3M}{h}\left\{1 - \frac{4(r - r_0)^2}{h^2}\right\} \tag{9.8}$$

and

$$\frac{\partial^2 \chi}{\partial r^2}\ (= \sigma_{\theta\theta}) = \frac{24M}{h^3}\ (r - r_0) \tag{9.9}$$

varying linearly through the thickness between the extreme values

$$\sigma_{\theta\theta}{}^{maximum} = 12\frac{M}{h^2} \text{ at } r = r_0 - \frac{h}{2}$$

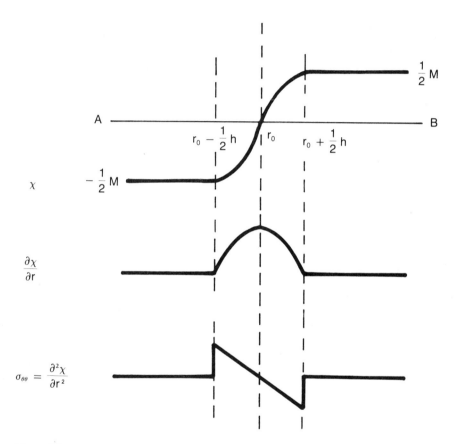

Figure 9.6. Variation through the thickness at AB in Figure 9.5 of χ, and its first and second differentials.

and

$$\sigma_{\theta\theta}{}^{minimum} = -12\frac{M}{h^2} \text{ at } r = r_0 + \frac{h}{2}$$

Equations (9.7), (9.8), (9.9) are shown graphically in Figure 9.6.

At $(r - r_0) = 0$, that is across the mid-plane, the χ surface is conical with, from Equation (9.8), $\partial\chi/\partial r = 3(M/h)$.

On such a surface, in the direction x orthogonal to the radius, $(\partial^2\chi/\partial x^2)$ $(= \sigma_{rr})$ is non-zero,* and equal to

$$\frac{1}{r}\frac{\partial\chi}{\partial r} = \frac{3M}{hr_0}$$

*For a conical χ surface,

$$\chi = ar$$

$$= a(x^2 + y^2)^{1/2}$$

$$\frac{\partial\chi}{\partial x} = a(x^2 + y^2)^{-1/2}x; \qquad \frac{\partial\chi}{\partial y} = a(x^2 + y^2)^{-1/2}y$$

$$\sigma_{yy} = \frac{\partial^2\chi}{\partial x^2} = -a(x^2 + y^2)^{-3/2}x^2 + a(x^2 + y^2)^{-1/2}$$

$$\sigma_{xx} = \frac{\partial^2\chi}{\partial y^2} = -a(x^2 + y^2)^{-3/2}y^2 + a(x^2 + y^2)^{-1/2}$$

$$-\sigma_{xy} = \frac{\partial^2\chi}{\partial x\partial y} = -a(x^2 + y^2)^{-3/2}xy$$

(9.10)

At $x = 0$,

$$\sigma_{rr} = \sigma_{yy} = \frac{\partial^2\chi}{\partial x^2}$$

$$= a(x^2 + y^2)^{-1/2}$$

$$= \frac{1}{y}\frac{\partial\chi}{\partial y}$$

$$= \frac{1}{r}\frac{\partial\chi}{\partial r}$$

(9.11)

and

$$\sigma_{\theta\theta} = \sigma_{xx} = \frac{\partial^2\chi}{\partial y^2} = 0,$$

$$\sigma_{\theta r} = \sigma_{xy} = -\frac{\partial^2\chi}{\partial x\partial y} = 0$$

which is $(h/4r_0)$ times $\sigma_{\theta\theta}^{maximum}$. That is, the ratio of this radial stress to the maximum longitudinal stress in a square cross-section curved beam of thickness 10 mm and inner radius 50 mm is

$$\frac{h}{4r_0} = \frac{10 \ (\text{mm})}{4 \times 5.5 \ (\text{mm})} = \frac{1}{22} = 4.55\%$$

This, however, is not the maximum value of $\sigma_{rr} = (1/r)(\partial\chi/\partial r)$, which occurs at somewhat smaller radius:

$$\frac{1}{r}\frac{\partial\chi}{\partial r} = \frac{3M}{h}\left\{1 - 4\frac{(r - r_0)^2}{h^2}\right\}\frac{1}{r}$$

$$\frac{d}{dr}\left(\frac{1}{r}\frac{\partial\chi}{\partial r}\right) = \frac{3M}{h}\left\{-\frac{8(r - r_0)}{h^2}\frac{1}{r} - \left(1 - \frac{4(r - r_0)^2}{h^2}\right)\frac{1}{r^2}\right\}$$

$$= \frac{3M}{h^3 r^2}\{-8(r - r_0)^2 - 8(r - r_0)r_0 - h^2 + 4(r - r_0)^2\}$$

$$= 0$$

making $\sigma_{rr}^{maximum}$ when

$$(r - r_0)^2 + 2r_0(r - r_0) - \frac{h^2}{4} = 0$$

$$(r - r_0) = r_0\left(1 \pm \sqrt{1 + \frac{h^2}{4r_0^2}}\right)$$

$$\cong r_0\left\{1 \pm \left(1 + \frac{h^2}{8r_0^2}\right)\right\}$$

The negative root is the one required, giving

$$(r - r_0) = -\frac{h^2}{8r_0} \tag{9.12}$$

This is $-h^2/44 = -0.23$ mm in the case considered ($r_0 = 55$ mm, $h = 10$ mm) making the radius for maximum radial stress 54.77 mm. Hence

$$\sigma_{rr}^{maximum} = \frac{3M}{h}\left\{1 - 4\left(\frac{h}{8r_0}\right)^2\right\}\left(r_0 - \frac{h^2}{8r_0}\right)^{-1}$$

$$\cong \frac{3M}{hr_0}\left\{ 1 - \frac{h^2}{16r_0^2} + \frac{h^2}{8r_0^2} \right\} \tag{9.13}$$

$$= \frac{3M}{hr_0}\left\{ 1 + \frac{h^2}{16r_0^2} \right\}$$

For the case considered, this makes a negligible fractional increase of about $1/500$ to $\sigma_{\theta\theta}{}^{maximum}$.

In this, and similar problems, it is important to note that it is anisotropy of strength, *not* anisotropy of elastic properties, which is significant.

4. Construct the Airy stress function in the air map for equal compression along diagonals of the confined cell formed by the arrangement of interfitting G-shaped anvils shown in Figure 2.12. Where in the anvils is the bending moment likely to be a maximum?

The G-shaped elements are drawn with heavy lines in Figure 9.7.

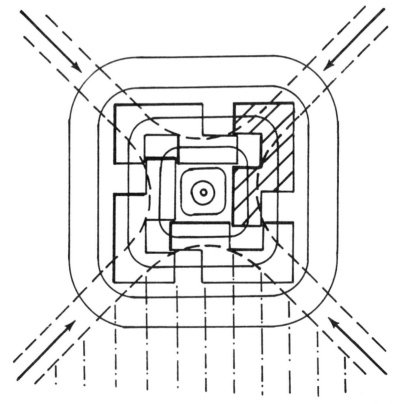

Figure 9.7. Airy stress function in the air for two-dimensional compression molding (cf. Figure 2.12).

In the center of the confined cell we have uniform pressure in two dimensions. $\nabla^2\chi = 0$ so the χ-surface here is a paraboloid, represented in the figure by the concentric circular contours of constant χ. Near the corners of the confined cell, the paraboloid will be rounded, i.e., the χ-surface changes to a parabolic pyramid close to the platen surfaces.

Between the lines of thrust, broken lines in Figure 9.7, we have membrane deformation, i.e., small amplitude bending. The contours of constant χ for this deformation are represented, in the lower quadrant only, by dot-dash parallel lines.

The bending moment is a maximum in the lower throat region of the G-shaped anvils.

9.3 Interlaminar shear strength

Simple shear = pure shear + a rotation (see Chapter 3, Section 3.11). Naively, it might be thought that simple shear is generated between the plies in a two-ply laminate subjected to the in-plane ply forces F shown in Figure 9.8(a). In fact, these forces produce a complicated stress distribution. Consider in Figure 9.8(b) the inter-ply material (resin). It is subjected to some kind of boundary

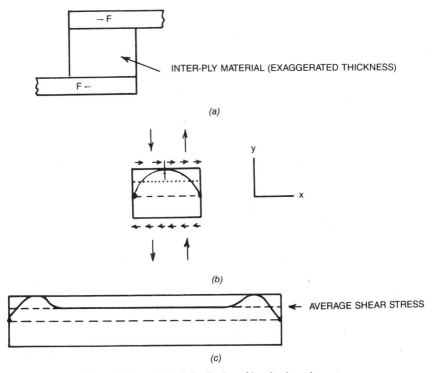

Figure 9.8. Longitudinal distribution of interlaminar shear stress.

stresses (surface tractions) represented by the small arrows. These alone would produce spin with angular acceleration. The fact that this does not occur means that other forces, represented by the larger arrows, are also present. To determine the directions of the latter, note the symmetry requirement that, after rotation of the specimen by 180° about the z-axis, the shear stresses must not have changed sign. Existence of the latter forces means that, in the upper one quarter thickness for example, there is a distribution of pressure which changes sign somewhere in the middle—across a vertical boundary represented by the vertical line.

Consider, now, the central horizontal plane through the resin layer—the broken line in Figure 9.8(b). Do we have uniform shear stress acting across this plane? The answer is no because at the free surface we have no tractions acting, so here the shear stress, indicated by the two dots at the ends of the broken line, is zero. (This is a general result—a free surface cannot have a shear stress acting at right angles to it.) So, what is the variation of shear stress τ across the central horizontal plane? We can sketch in the average shear stress = F/area of bonded surface. This is the horizontal dotted line in Figure 9.8(b). And we can, as a first attempt, guess at the parabolic distribution also sketched in Figure 9.8(b). This, in fact, is about right but a difference arises when we consider the more realistic shape for the inter-ply layer shown in Figure 9.8(c).

What we have now is simple shear in the middle, but the boundaries (the edges of the plies) complicate matters. The two bumps in the distribution of mid-plane shear stress arise because the area under the line corresponding to the average shear stress has to equal the area above that line. Details of the distribution of shear were first calculated by C. E. Inglis.[39] Novel confirmation of Inglis' result was obtained by L. Bragg and W. M. Lomer[40] in their experimental study of homogeneous nucleation of dislocations in sheared bubble rafts. Bragg and Lomer observed the creation of dislocation pairs at locations corresponding to the two maxima in wide rafts, and in the middle, corresponding to the single maximum shear stress, in narrow rafts.

A complete treatment of how elastic anisotropy affects the detailed distribution of stress across a bent beam, with edge effects included, is beyond the scope of this book. At the edges of wide beams there arise stress concentrations due, essentially, to Poisson's ratio effects; these effects would be zero if the Poisson's ratios were zero. Knowledge of a single interlaminar shear strength is not enough to account for these effects. It is essential to consider two interlaminar shear strengths, one for shear along and the other for shear transverse to the fibers (cf. Chapter 4, Sections 4.2 and 4.3, and Figure 4.7). Delamination by transverse shearing near the edges can be a significant mode of failure commencement in the bending of a wide laminated beam.

[39]C. E. Inglis. "Stress Distribution in a Rectangular Plate Having Two Opposing Edges Sheared in Opposite Directions," *Proc. Roy. Soc.*, A 103:598–610 (1923).
[40]L. Bragg and W. M. Lomer. "A Dynamical Model of a Crystal Structure, II," *Proc. Roy. Soc.*, A 196:171–181 (1949).

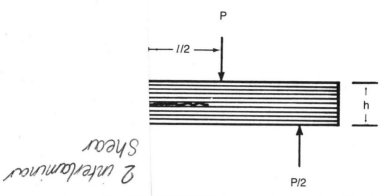

Figure 9.9. ASTM D-2344 short beam shear test-piece. $l/h = 4$.

9.4 Short beam shear test

There is a widely used American Society for Testing Materials (ASTM) three-point bend test known as ASTM D-30 standards # D 2344 "Apparent Interlaminar Shear Strength of Parallel Fiber Composites by Short Beam Method." It is claimed that the dimensions of the test-piece, see Figure 9.9, are so specified as to induce interlaminar shear failure along the neutral axis before failure in the tensioned surface of the beam.

The apparent interlaminar shear strength τ is taken to be the maximum shear stress predicted by classical elasticity theory on the neutral axis of an isotropic material, that is

$$\tau = \frac{3P}{4bh} \tag{9.14}$$

where P is the center load and b is the width of the beam. Implicit in the derivation of Equation (9.14) is that the maximum bending moment is $Pl/4$, where l is the distance between the outer supports. $Pl/4$ for the maximum bending moment is valid only for the most rigid of test-pieces such as resin bonded close-packed fiberglass for which the fiber volume fraction

$$\eta = \frac{\pi}{2\sqrt{3}}$$

that is for a test-piece containing 90.69% by volume of glass. For flexible test-pieces, i.e., test-pieces with lower fiber volume fractions, the maximum bending moment exceeds $Pl/4$ as the following analysis demonstrates.

The Airy stress function in the air is sketched in Figure 9.10; there are three areas of air space, labelled a, b and c.

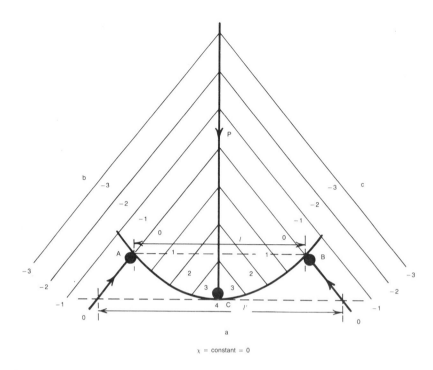

Figure 9.10. Airy stress function in the air χ-diagram for deep three-point bending (χ-scale arbitrary).

The line of action of the center load (compressive) delineates a ridge-fold in the χ surface, so the magnitude of χ decreases linearly as we move out on any straight line at right angles to the line of action of the center load. The lines of action at the outer supports are orthogonal to the specimen and each of these must also delineate a ridge-fold. Hence the contours (χ-scale arbitrary) are as shown.

It can be seen from this diagram that the maximum bending moment, maximum $\nabla\chi$ — see question 2, occurring at C, exceeds $Pl/4$, being equal to $Pl'/4$ where l' is the distance between intercepts of normals to the beam at A and B with the horizontal through C. In order to determine l' experimentally, it would be necessary to make provision for determining the positions of the lines of contact between the specimen and the knife-edges or cylinders across which the load is applied. One method would be transmission of ultrasound.

The upshot is that use of Equation (9.14) partially masks the dependence of interlaminar shear strength on fiber volume fraction. $\tau = 70$–100 MN m^{-2} for glass fiber and carbon fiber reinforced resins containing fiber volume fractions $\eta = 1/2$. Typically quoted in the literature are apparent reductions in interlaminar shear strength of $\cong 50\%$ when the fiber volume fraction is increased from $\eta = 1/2$ to $\eta = 3/4$. Much bigger reductions in τ with increasing fiber content would be revealed if a corrected Equation (9.14) were used for the more flexible low fiber volume fraction test-pieces.

The ASTM short beam test takes no cognizance of the above analysis, or of the existence of two interlaminar shear strengths (see Section 9.3). In addition, failure is often complex. Thick composites (50 plies, for example), often fail in three-point bending by creasing [kinking – see Chapter 8, Section 8.7 and Section 8.9, question 1 (b)] near the center load. At best, ASTM D-2344 is useful only for comparative tests on thin composites of the same fiber volume fraction, and then only if the mode of failure is carefully recorded.

9.5 F. C. Frank's seismic fault model for delamination

The model described in this section, and developed in Sections 9.6 and 9.7 is taken from F. C. Frank's[26] publication on seismic sources. The physics of propagation of shear cracks in the resin-rich material between plies in a laminate is not unlike that for the propagation of shear cracks (earthquakes) in the crust and upper mantle of the Earth.

Let us idealize the region of delamination as an ellipsoid with semi-axes $a > b > c$. Its extent is elastically limited by its surroundings and, assuming no volume change, the main effect of shear stress is an enhancement of the shear strain on a plane parallel to the plane of flattening of the ellipsoid and an enhancement of the corresponding shear stress around the edge.

One way of describing the stress system around such regions of delamination is to regard it as equivalent to a system of concentric dislocations. The stress due to an edge dislocation is larger than the stress due to a screw dislocation by a factor $1/(1 - \nu)$, where ν is Poisson's ratio (say 0.38 for epoxy resin, so that this factor is 1.61). By way of compensation for this greater stress, the screw segments of dislocation are able to crowd together towards the edge of the delaminating region, reinforcing each other's stress field more effectively. The effective frictional forces restraining shear will be different for edge and screw segments, the upshot of which is that delamination will tend to grow more rapidly in one of the two directions, along or across the direction of delamination, so that the typical region of delamination from a single origin will be elongated. The geometrical constraints demand that the delamination be straight in the direction of shear but allow some bending of its plane in the transverse direction; screw dislocations are free to change their glide plane, edge dislocations are not.

What we have described here is very similar to the development of Lüder's bands in polycrystalline iron or in any material which becomes weaker after a certain small strain; the regions of concentrated deformation grow in directions determined by the stress conditions and not by the previous structure of the material. The mechanism is significantly different, but the factors of geometrical constraint operate in very much the same way.

The shape of an individual region of delamination in a plane is probably well described as an ellipse with semi-axes $a > b$. Taking into account the transverse cross-section, extension of the ellipse to an ellipsoid is no more than a rough description. However, all that really matters is the radius of curvature (ϱ) at the edge of the delamination. The stress at the edge is increased from its distant value (τ_0) by a factor of about $2\sqrt{b/\varrho}$, Equation (4.5).

The strain energy released by the delaminated region, assuming total relaxation of shear stress within it, is about

$$\frac{ab^2}{\mu}\tau_0^2 \tag{9.15}$$

where μ is the shear modulus of the laminate. If, as is likely, the value of ϱ remains approximately constant as the delaminating region grows, the stress concentration near its edge becomes steadily greater, causing accelerated growth. Eventually, the stress concentration will become equal to the interlaminar critical shear strength, namely when

$$\frac{b}{\varrho} = \left(\frac{\tau_{critical}}{2\tau_0}\right)^2 \tag{9.16}$$

The delamination can now propagate.

9.6 Transition to fast delamination

At this stage, it is necessary to consider the possibility, for example in thermoplastic matrix laminates, that delamination becomes catastrophic because the work of plastic deformation under concentrated stress at the edge of the delaminating region creates enough melting to destroy the strength, advancing the position of stress concentration without weakening it and so allowing a rapid propagation of failure.

Treating the problem after the manner of Griffith,[7] the elastic energy release at the delamination ($W_e \cong ab^2\tau_0^2/\mu$) increases by

$$dW_e \cong (2ab\,db + b^2\,da)\tau_0^2/\mu \tag{9.17}$$

for growth increments db, da, while the area of delamination increases by $dA = \pi(a\,db + b\,da)$. Assuming that the residual strength within the delaminated region is negligible and that most of the work is converted to heat at its growing edge, the energy so deposited per increment of area varies from $b\tau_0^2/\pi\mu$ for end growth to $2b\tau_0^2/\pi\mu$ for side growth. Assuming $\tau = 10$ MN m^{-2}, $\mu = 100$ GN m^{-2} and a latent heat of melting of 1.6×10^9 J m^{-3}, $2b\tau_0^2/\pi\mu$ is sufficient to create a molten layer of thickness $4 \times 10^{-7}b$. This becomes $\cong 0.5$

μm when $b = 1$ m and may be sufficient for virtual destruction of the shear strength of the interlaminar material, allowing catastrophic delamination.

9.7 Junction of delaminated regions

Since a single delaminating zone must be large before it can undergo the transition to catastrophically rapid growth, it is likely to meet other delaminating zones before the transition occurs. Growth by way of junction of delaminated regions certainly leads to large and rapid releases of elastically stored energy. Where the edges of the zones approach, their stress concentrations reinforce each other so that their rates of growth towards each other will be enhanced and will accelerate as junction is approached.

The process of junction is sketched in Figure 9.11 for the case of two delaminating zones meeting side by side in the same plane. In this figure, each region of delamination is represented by concentric loops of dislocation. The relative displacement of the opposite surfaces of the delaminated region at any point is proportional to the number of dislocation lines that have passed that point. The dislocation line analogy affords qualitative visualization not only of the displacements but also of the stress concentrations. Dislocations of opposite sign attract and dislocations of the same sign repel each other with forces inversely proportional to the distance between them. Thus, there is high stress where the dislocation lines crowd together.

Evidently, there is a very high concentration of stress produced by the coalescence of two dislocation loops at the re-entrant points in the periphery of the delamination zone sketched in Figure 9.11(b). Although the delamination zone after coalescence [Figure 9.11(c)] is still smaller than the size needed for catastrophic growth by shear melting in thermoplastics—see Section 9.6, elimination of the re-entrants when making the transition from Figure 9.11(b) to Figure 9.11(c)

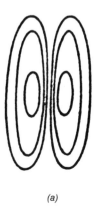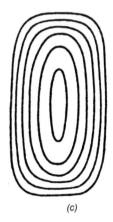

(a) (b) (c)

Figure 9.11. Showing the junction of two delaminations meeting side by side in the same plane.

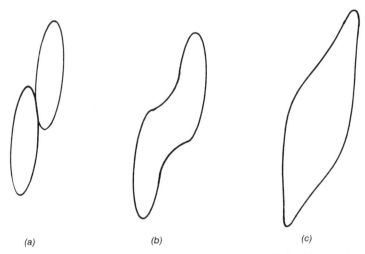

(a)   *(b)*   *(c)*

Figure 9.12. Union of two delaminations which are not exactly side by side or end to end.

may well occur by melting. The energy released in the transition from Figure 9.11(b) to (c) is approximately

$$\{a(2b)^2 - 2ab^2\}\tau_0^2/\mu = 2ab^2\tau_0^2/\mu \tag{9.18}$$

In estimating the energy released by coalescence of delaminating regions, both the separate regions before union and the single region produced by their junction have been approximated as ellipses. This approximation predicts zero release of energy if two elongated regions of delamination join end to end. In reality, end-wise union no doubt gives rise to a relatively small release of energy so zero might be an appropriate approximation.

When the two uniting regions of delamination lie neither exactly side by side nor exactly end to end (see Figure 9.12), rapid release of an amount of energy proportional to the overlapping length in the transition from Figure 9.12(a) to

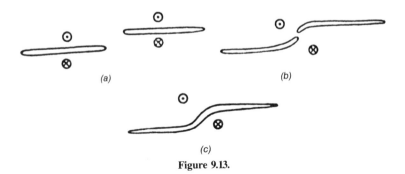

(a)   *(b)*

(c)

Figure 9.13.

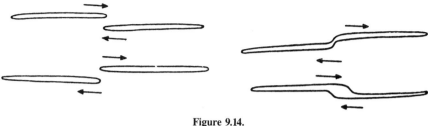

Figure 9.14.

Figure 9.12(b) is anticipated. A further rapid release of energy is expected to accompany elimination of the re-entrant parts of the boundary in the transition to Figure 9.12(c).

No indication of the direction of shear is given in Figures 9.11 and 9.12, but it is of little importance compared with the direction of elongation of the delaminating regions. The direction of shear only makes changes by a factor between $(1 - \nu)$ and 1 in the dislocation stresses and in the energy released during the coalescence.

The direction of shear has more important consequences for the coalescence of delaminating regions that are not coplanar (see Figure 9.13). The stress concentration in the intervening material, where the edges of delaminations propagate past each other, will, if fiber orientation and/or fiber breakages permit, cause coalescence to take place. The result is a non-planar delamination with an elongated step. An example of this phenomenon is given in Figure 9.13. The step does not impede further delamination if, as in Figure 9.13, it is parallel to the direction of shear, but the step does impede further delamination if it is across the shear direction as in Figure 9.14.

9.8 Out-of-plane stresses

Delamination is usually initiated by out-of-plane stresses of which there exist several sources including the free-edge (discussed in Chapter 4, Section 4.3), through-the-thickness holes in laminates subjected to in-plane compression [cf. Chapter 4, Section 4.4, parts (ii) and (iv)], ply-drops incorporated to taper the laminate thickness (the edge of an internally terminated ply creates a line singularity that is analogous to an edge dislocation in a crystal), adhesively bonded joints (see Chapter 11), and outward buckling of delaminated plies under in-plane compression (see Section 8.7 and Figure 8.30). These five sources of out-of-plane stress are summarized in Figure 9.15.

9.9 Anisotropy of plasticity in
metal matrix composites

Plastic deformation of the metal in metal matrix composites is impeded by the fibers which can deform only elastically. This impediment to plastic deformation

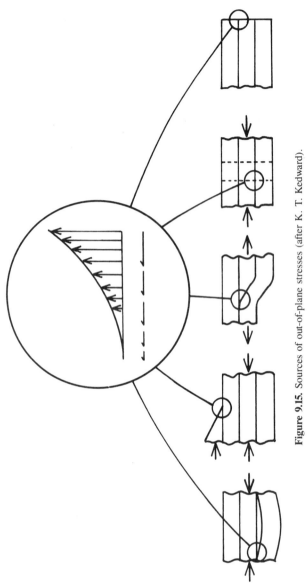

Figure 9.15. Sources of out-of-plane stresses (after K. T. Kedward).

by the fibers comes from three sources:

(1) *The presence of a back-stress.* The plastic deformation of the matrix is characterized by a stress-free strain since, were the fibers not present, it would merely change the shape of the holes in which they sit without generating any stress in the matrix. The fact that the fibers are present means that elastic stresses and strains are created in both matrix and fibers, and these stresses and strains can be calculated using Eshelby's[10] theory for the elastic field of an ellipsoidal inclusion.

L. M. Brown and D. R. Clarke[41] find that, for primary slip in a single crystal: (a) the internal stress is a maximum when the fibers are parallel to the axis of principal extension of the composite, (b) if the fibers are either parallel to the slip direction, or perpendicular to the slip plane, or parallel to the slip plane and perpendicular to the slip direction, then the internal stress is one of simple shear proportional to the overall plastic shear strain of the composite, and (c) for all other fiber orientations the internal stress is not simple shear; it involves dilation — the composite is plastically anisotropic — the components of internal stress are not in proportion to the components of overall plastic strain. Of special note for the general case (c) is that the fibers are axially compressed when the composite is subjected to overall uniaxial tension and vice versa.

(2) *Shortening of the dislocation glide distance.* When a gliding dislocation encounters an array of fibers inclined to the slip plane, it bows out between them and, given sufficient applied stress, will loop around the fibers and escape by way of mutual annihilation where segments of opposite sign meet, leaving so-called Orowan loops of dislocation around each fiber. This process is illustrated in Figure 9.16.

The Orowan loops repel approach of the next dislocation gliding on the same slip plane and thereby effect a decrease in the spacing between the obstacles to slip. Hence, the second gliding dislocation is obliged to bend with larger curvature, than was the first, in order to loop between the fibers. The result is increased resistance to plastic deformation of the matrix.

(3) *Introduction of non-gliding dislocations which inhibit the movement of dislocations which can glide.* Differential contraction between the fibers and matrix during cooling from the temperature at which the composite was fabricated is mainly accommodated by creating dislocations in the matrix. Those which lie more or less parallel to gliding dislocations but are immobile, on account of pinning by impurity atoms for example, will interact elastically, and those which lie more or less perpendicular to active slip planes will have to be intersected by gliding dislocations. Both orientations of pre-existing dislocations therefore give rise to hardening.

[41]L. M. Brown and D. R. Clarke. "The Work-Hardening of Fibrous Composites with Particular Reference to the Copper-Tungsten System," *Acta. Met.*, 25:563–570 (1977).

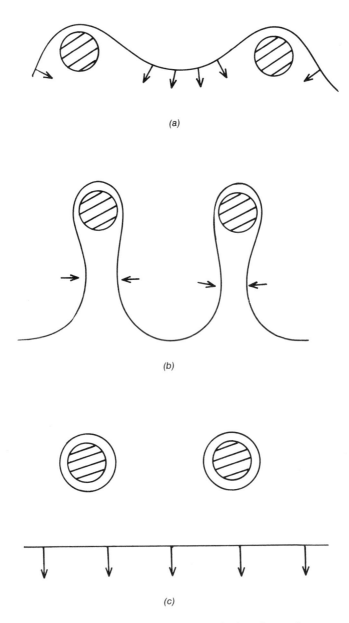

Figure 9.16. (a) A dislocation, gliding from top to bottom, begins to bow out between two fibers. (b) The segments of dislocation line with opposite sign begin to attract each other (arrowed). (c) The dislocation escapes leaving an Orowan loop around each fiber.

9.10 Fatigue damage accumulation

Laminates subjected to in-plane uniaxial tension-tension fatigue loading undergo well distributed* matrix cracking and delamination. Matrix cracks appear first in the 90° plies, then in the 45° plies. Delamination commences between 45°/90° and is initiated at the free-edges. Later on, delaminations appear at locations where matrix cracks in adjacent plies pass each other. Towards the end of the fatigue life-time, matrix cracks develop even in 0° plies. The matrix cracks, especially those in the 90° plies, are observed to develop into a uniform array typically a couple of millimeters apart.

A convenient way to monitor damage accumulation is to measure the modulus and, by defining a damage parameter in terms of modulus, it is possible to come up with empirical relationships between damage rate and mean stress. One such relationship, due to A. Poursartip, M. F. Ashby and P. W. R. Beaumont[42] is

$$\frac{dD}{dN} \propto (\sigma_m)^{2.71}$$

9.11 Further examples

(1) The pull required to fully tension a graphite fiber/epoxy resin laminated cross-bow is 200 N. Calculate the maximum shear force in the bar at distance 0.5 m from one end. Assume that the tensioned string makes angles of 60° with the stem.

(2) Use Equation (9.18) to estimate the energy released during elimination of the re-entrants created by union of two delaminations, each measuring 100 mm long by 50 mm wide, meeting side by side in the same plane in a quasi-isotropic laminate with in-plane shear modulus 100 GN m^{-2} subjected to an in-plane shear stress of 10 MN m^{-2}.

(3) Sketch the Airy stress function in the air for four-point bending and show that the bending moment is uniform between the central knife-edges.

(4) Figure 9.17 shows a reverse bend fatigue test piece. It is tapered along its gage length so as to make the bending moment uniform when gripped at its wide end and flexed by displacement at its narrow end. Construct the Airy stress function in the air for this test-piece and estimate the taper required to make the bending moment uniform within the gage length.

(5) It can be shown that in a state of plane strain (and also plane stress) in an elastic body, the Airy stress function χ satisfies the equation:

$$\nabla^4(\chi) = 0$$

---

*Contrast the localised nature of defects in more conventional materials such as metals.

[42]A. Poursartip, M. F. Ashby and P. W. R. Beaumont. "The Fatigue Damage Mechanics of Fibrous Laminates," in *Polymer NDE*, edited by K. H. G. Ashbee, Lancaster, PA:Technomic Publishing Co., Inc., pp. 250–260 (1985).

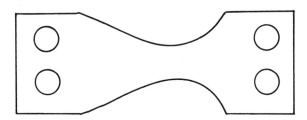

Figure 9.17. Reverse bend fatigue test piece.

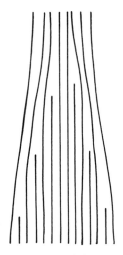

Figure 9.18.

where

$$\nabla^2 = \frac{\partial^2}{\partial x^2} + \frac{\partial^2}{\partial y^2}$$

Examine the function $\chi = ay^3$, and hence determine the deformation that it represents (University of Bristol Examination, 1973).

(6) What problem is solved exactly by

$$\chi = -\frac{F}{d^3} xy^2 (3d - 2y)$$

applied to the region

$$0 < y < d; x > 0?$$

What problems would be solved approximately by the above expression (University of Bristol Examination, 1972)?

(7) (Library project) Figure 9.18 illustrates one method whereby "ply-drops" are used to effect a change in thickness of a laminate, as might be required for the vertical stabilizer on an aircraft. Geometrically, the ply-drops, that is the edges of the terminated plies, constitute edge dislocations in the laminate. Develop this analogy by considering the elastic field of an edge dislocation in order to describe the local bending and its associated stresses in the vicinity of a ply-drop.

(8) When manufactured from light alloy, the tension skin of an aircraft wing needs to be thicker than the compression skin, vice versa when manufactured from a laminate. Explain.

10

Environmental degradation

In normal use, most fiber reinforced materials are exposed only to the Earth's atmosphere. For some applications of composites this exposure may occur at high speeds in which case erosion may be the dominant mechanism of environmental degradation. Erosion of composites is discussed in Chapter 8, Section 8.8 and will not be further discussed here. Neither will degradation in other hostile environments and here we include extreme temperature (hot or cold), strong acids and strong alkalis.

The average chemical composition, excluding water vapor, of the Earth's atmosphere is nitrogen 78%, oxygen 21%, argon <1%, carbon dioxide 0.03% by volume, prolonged exposure to which gives rise to chemical and physical changes in each of polymer matrix and metal matrix composites.

The polymers absorb water, the accommodation of which produces dimensional changes and hence internal stress. The absorbed water may also seek out and dissolve water soluble inclusions, thereby creating pockets of concentrated solutions which, by either chemical or physical means lead to loss of mechanical strength of the composite. The mechanisms of degradation range from static fatigue of fibers, to creation of osmotic pressure filled interfacial cracks, to shattering due to crystallization of water. The consequent losses in mechanical properties can be, for carbon fiber/epoxy resin composites, of the order of 50% for interlaminar shear strength, 50% for compressive strength and 10% for tensile strength.

Exposure to sunlight causes physical degradation of polymers but this is confined to surface layers down to about 10 μm depth, 10 μm being a typical thickness of resin at which the intensity of ultraviolet light is reduced to one half its incident value [see Chapter 12, Equation (12.8)]. The nature of the degradation by ultraviolet radiation involves cross-linking leading to shrinkage and embrittlement. Although it spoils the aesthetic appearance of the surface, ultraviolet degradation usually imparts no significant damage to the fiber reinforced material and will not, therefore, be pursued here.

Environmental degradation of metal matrix composites usually takes the form of metal corrosion. Almost all oxides are less dense than the parent metal so when oxidation occurs in a confined space, at the metal/fiber interface for example, the volume expansion works against a confining pressure. This problem and its consequences is well described in metallurgical text-books.

This chapter will, therefore, concentrate on the degradation of resin matrix composites by solvents, mainly water but also organic solvents including petroleum products.

10.1 Swelling stresses due to non-uniform water uptake

The polymers, as a class of materials, absorb fluids and in doing so they undergo dimensional change. In the case of water absorption the dimensional change is usually a volume expansion* which, for epoxy resins amounts to a few per cent at saturation. Absorbed water is rarely uniformly distributed and, because of this, there is usually a distribution of internal stress associated with water uptake.

Consider in Figure 10.1 a parallel-sided slab of resin exposed to an aqueous environment. The upper row of sketches shows the progress with time of water uptake. At the start, the water concentration (c) is zero across the thickness and the slab is stress free. As soon as exposure commences, second sketch from the left, the surface layers become saturated with water but the deeper layers are still dry. The surface swelling associated with the diffused water generates compression parallel to the slab and this requires that tension be simultaneously generated in the deeper layers. This state of stress, compression parallel to the slab in the layers containing water and tension in other layers, is represented by the sketch below the water distribution sketch. As water uptake proceeds, the stress parallel to the plate is redistributed until it is zero all across the thickness of the slab at uniformly saturated water uptake (far right sketch).

Stresses of opposite sense are created during drying out of a slab uniformly saturated with water. Starting with a uniformly swollen and therefore stress-free slab, left-hand sketch in the upper row of Figure 10.2, drying-out of the surface tensions these layers and at the same time introduces compression parallel to the plate in deeper layers. As drying continues, this distribution of stress changes, ending with a stress-free state when all the water has been removed (far right-hand sketches in Figure 10.2).

Now consider water uptake and its associated swelling by the resin in a uniaxial fiber reinforced composite. Between three closely packed parallel fibers, the resin forms a fillet with cross-section a circular triangle. This is shown in heavy outline in the upper diagram in Figure 10.3. Water from the environment enters the fillet mainly at its ends so, if free to do so, the ends would swell to larger cross-

---

*An important exception is the polyester resins which, although swollen by small water uptake, actually undergo net shrinkage after prolonged exposure to water.

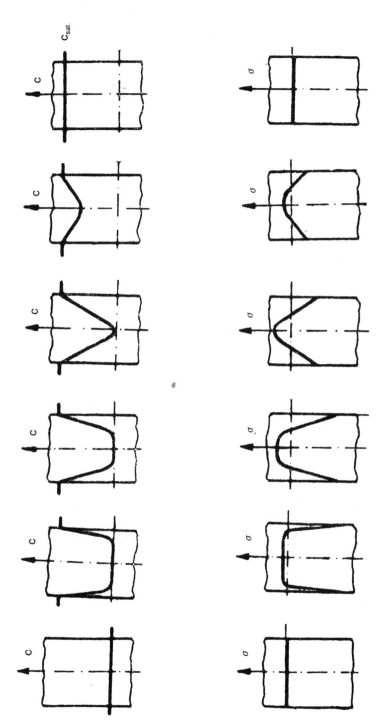

Figure 10.1. Water uptake through the thickness (upper row) and associated in-plane stress distribution (lower row) for a slab of neat resin, as a function of time of exposure to aqueous environment (after D. Putz).

269

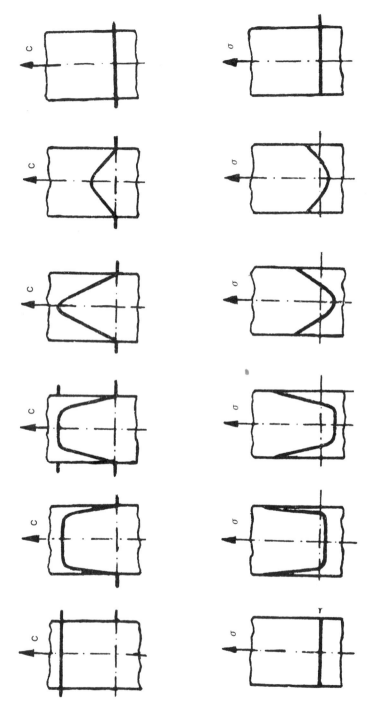

Figure 10.2.

270

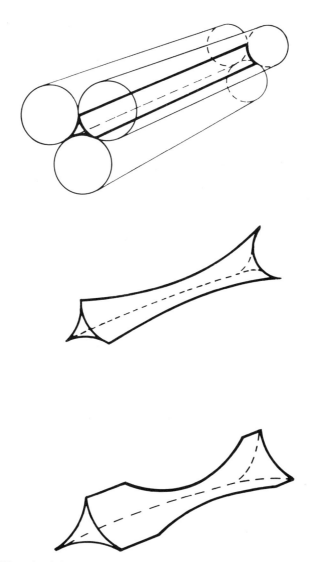

Figure 10.3. Dimensional changes that the resin between fibers would like to undergo during water uptake.

section as indicated in the center diagram in Figure 10.3. If water uptake at the ends continues to saturation, the free geometry for the swollen fillet would look like that sketched in the lower diagram in Figure 10.3. However, the resin is not free to undergo these changes in shape; it is constrained by the bending stiffness of the fibers, backed up by the rest of the composite, not to swell. As a consequence, radial compression is generated at the ends of the resin fillet at the onset of water uptake and, at the same time, radial tension is generated at larger distances from the ends.

If the fillet of resin between three fibers were uniformly swollen as a consequence of water uptake, the fibers would be tensioned* and displaced away from each other without bending, and the overall deformation would be opposite to that produced by uniform resin shrinkage during curing. In this case, the swelling could be treated in the same way as thermal expansion.

Assuming that tensile stress is relieved by the presence of water molecules, the internal tensile stress generated as described above by inhomogeneous water uptake is expected to attract absorbed water and hence give rise to both enhanced water migration rates and super-saturated water uptake.

An estimate of the stress required for stress induced drift of water molecules to be a significant enhancement of diffusion may be made as follows. The respective fluxes are

$$j_{drift} = -\mu c \, \nabla \phi \tag{10.1}$$

where μ is the drift mobility, c is the water concentration and $\nabla \phi$ is the potential energy gradient arising from the tensile stress at the center of the specimen sketched in Figure 10.3, and

$$j_{diffusion} = -D \, \nabla c \tag{10.2}$$

where D is the water diffusion coefficient and ∇c is the water concentration gradient.

$$\frac{j_{drift}}{j_{diffusion}} = \frac{\mu c \nabla \phi}{D \nabla c} = \frac{c}{\nabla c} \cdot \frac{\nabla \phi}{kT}$$

using the Nernst-Einstein relationship $\mu = D/kT$. The dimensions are meters for $c/\nabla c$ and J m^{-1} for $\nabla \phi$. Hence,

$$\frac{c}{\nabla c} \, \nabla \phi$$

---

*In model composites containing very low fiber volume fractions, the axial tension due to resin swelling can be sufficiently large to fracture the fiber.

is a quantity of energy, say ΔE, i.e.,

$$\frac{j_{drift}}{j_{diffusion}} \cong \frac{\Delta E}{kT} \tag{10.3}$$

At room temperature

$$kT \cong \frac{1}{40} \, (eV)$$

$$= \frac{1.6 \times 10^{-19}}{40} \, (J)$$

Hence, for the stress induced flux to be say, 10% of the normal diffusion flux at room temperature, ΔE would have to be of the order of 4×10^{-22} joules.

Assuming that water molecules migrate singly and cause swelling ΔV per molecule equal in magnitude to the natural volume of the water molecule, where

$$\Delta V = \frac{18(gm)/1(gm/cm^3)}{6 \times 10^{23}}$$

$$= 3 \times 10^{-23} \, cm^3$$

$$= 3 \times 10^{-29} \, m^3$$

then $\Delta E = 4 \times 10^{-22} \, J = p \, \Delta V$ and

$$p = \frac{4 \times 10^{-22}(J)}{3 \times 10^{-29}(m^3)}$$

$$= 1.3 \times 10^7 (N/m^2) \text{ since } N = (J/m)$$

$$= 13 \, MN \, m^{-2}$$

which is a modest value for internal stresses generated by inhomogeneous swelling in composite materials.

10.2 Osmosis

Abbé Nollet, in 1748, found that when a bladder full of alcohol is immersed in water, the water enters the bladder more rapidly than the alcohol escapes, so the

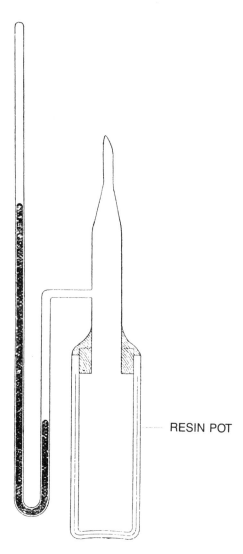

Figure 10.4.

RESIN POT

bladder swells out and almost bursts. If on the other hand, a bladder containing water is placed in alcohol the bladder shrinks. The passage of liquids through membranes is called osmosis.

There are membranes which, while permeable to water, are impermeable to a large number of salts; these membranes are called semi-permeable. Polyester and epoxy resins are examples. If a resin pot fitted with a pressure gage, as in Figure 10.4 is filled with a dilute solution of a salt and is immersed in pure water,* water will flow into the pot and compress the air in the gage, the pressure in the pot increasing until a definite pressure is reached depending on the concentration of the solution. When this pressure is reached there is equilibrium, and there is no further increase in the volume of water in the pot.

Thus the flow of water through the membrane into the more concentrated solution can be prevented by applying to the solution a definite pressure; this pressure is called the osmotic pressure of the solution.

The work done when a volume v of water passes across a semi-permeable membrane from pure water into a solution where the osmotic pressure is P is equal to Pv. Suppose the solution is enclosed in a vertical tube closed at the lower end by a semi-permeable membrane, Figure 10.5. Then, when there is equilibrium the solution is at such a height in the tube that the pressure at the membrane due to the head of the solution is equal to the osmotic pressure. When the system is in equilibrium we know that the total work done during any small alteration of the system must be zero. Let this alteration be a volume v of water going through the semi-permeable membrane. This volume will raise the level of the solution, and the work done against gravity is the same as if a volume v of the solution were raised from the level of the membrane to that of the top of the liquid in the tube. Thus the work done against gravity is $vg\varrho h$, where h is the height of the solution in the tube and ϱ is the density of the solution. Since the pressure due to the head of solution is equal to the osmotic pressure, $g\varrho h = P$. Hence the work done against gravity by this alteration is Pv and, since the total work done must be zero, the work done on the liquid when it crosses the membrane must be $-Pv$.

The magnitudes of the osmotic pressures for different solutions were first measured by Pfeffer,[43] who found the remarkable result that, for dilute non-electrolytic solutions, the osmotic pressure is equal to the gaseous pressure which would be exerted by the molecules of the salt if they were in the gaseous state and occupied a volume equal to that of the solvent in which the salt is dissolved. Thus if 1 gram-equivalent of the salt were dissolved in a liter of water the osmotic pressure would be about 22 bars. Pfeffer's experiments also showed that, like the pressure of a gas, the osmotic pressure is proportional to the absolute

---

*A source of water at lower chemical potential than pure water, that is an aqueous solution, will generate smaller osmotic pressure. This explains why osmosis is less of a problem in saline than in fresh water environments.

[43]W. Pfeffer. *Osmotische Untersuchungen*, Verlag von Wilhelm Englemann, Leipzig (1877).

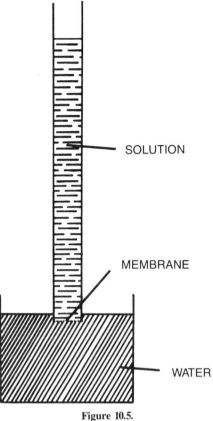

Figure 10.5.

temperature:

$$p = \frac{RT}{V} \tag{10.4}$$

usually referred to as Van't Hoff's law of osmotic pressure, where R is the gas constant, T is the absolute temperature and V is the volume of solution containing 1 g molecule of solute.

10.3 Osmotic pressure filled cracks[44]

Exposure to hot water of many fully cured general purpose polyester resins results in isolated disc-shaped cracks, usually misoriented with respect to each

[44]K. H. G. Ashbee, F. C. Frank and R. C. Wyatt. "Water Damage in Polyester Resins," *Proc. Roy. Soc.*, A300:415–419 (1967).

other and typically <1 mm in diameter. When formed, such cracks are filled with water (actually water solution) as may be demonstrated by the orientations of cracks for which total internal reflection occurs, see Figure 10.6. Scanning electron microscopy of the surfaces of such cracks, removed from specimens which have been dried after immersion in water, reveals the presence of solid deposits, Figure 10.7, and some of these deposits are large enough for electron probe X-ray micro-analysis, application of which suggests that they contain inorganic water solubles, potassium chloride for the deposit analyzed in Figure 10.8. Solutions of organic solutes must also exist. However, identification of organic material within deposits requires molecular information as well as atomic information, and molecular analysis of micron size samples is very much more difficult to accomplish.

A normal solution is one which contains the equivalent weight of a substance (in grams) dissolved in one liter of solution. The equivalent of a salt is the number of grams of it which, by reaction of an acid with a base, causes one gram of hydrogen to be replaced by a metal. Taking as example,

$$KOH + HCl = KCl + H_2O$$

a 1 N solution of KCl contains 1 gm mole $= 74.5$ gm KCl in 1 liter of water and this gives rise to an osmosis pressure of 22 bars at NTP.

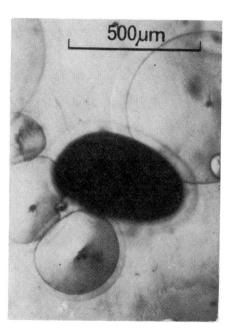

Figure 10.6. Transmission optical micrograph of disc-shaped cracks in neat polyester resin. Note the total internal reflection from the central (air-filled) portion of one of these cracks.

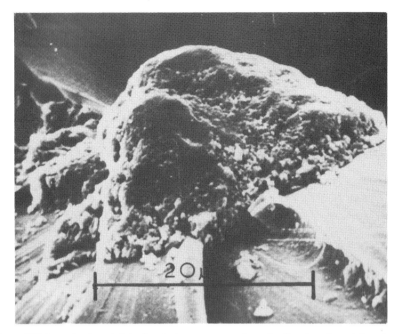

Figure 10.7. Impurity deposit left after drying, on one surface of a disc-shaped crack in polyester resin.[44]

Taking $100\,°C$ as our water uptake temperature, the osmotic pressure generated by a 1 N KCl solution is, according to Boyle's and Charles' laws combined

$$\frac{P_1 V_1}{T_1} = \frac{P_2 V_2}{T_2}$$

with $V_2 \cong V_1$

$$P_2 = \frac{P_1 T_2}{T_1}$$

$$= \frac{22(\text{bars}) \times 373\ (\text{K})}{293\ (\text{K})}$$

$$= 28\ \text{bars}$$

(10.5)

The hot wet strength of polyester resin is of the order of 40 MN m^{-2}. To get an osmotic pressure of 40 MN m^{-2} (400 bars) would require a $400/28 \cong 14$ N solution.

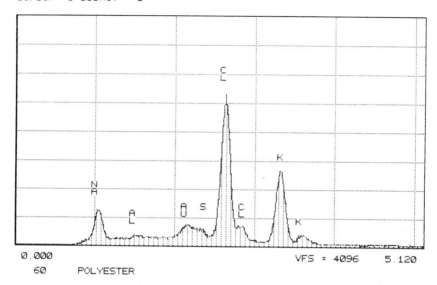

0.000   VFS = 4096   5.120
  60   POLYESTER

SETUP SSQ

SSQ: SSQ

SEMI-QUANTITATIVE ANALYSIS: POLYESTER
EL   NORM. K-RATIO

NA-K 0.13111 +- 0.00184
K -K 0.33167 +- 0.00269
CL-K 0.53721 +- 0.00315

ZAF CORRECTION   15.00 KV   50.00 Degs

No. of Iterations 2

----	K	[Z]	[A]	[F]	[ZAF]	ATOM.%	WT.%	
NA-K	0.131	0.940	1.529	0.997	1.434	24.23	16.75	*
K -K	0.331	1.024	1.148	1.000	1.175	29.63	34.73	
CL-K	0.537	1.004	1.031	0.978	1.014	46.13	48.52	

* - High Absorbance

Figure 10.8. Energy dispersive analysis of the deposit shown in Figure 10.7.

10.4 Osmotic pressures

When inorganic salts are dissolved in water, the osmotic pressure is greater than in a solution of equivalent concentration of organic salts. This increase is explained by Arrhenius as being due to ionic dissociation. Thus NaCl in solution dissociates into Na⁺ and Cl⁻ ions and, since by this dissociation the number of individual particles in unit volume is increased, the osmotic pressure (if it follows the gas laws) will also be increased.

Published measurements of osmotic pressures have been compiled in Landolt-Bornstein.[45] The data reported includes the following pressures for inorganic salts dissolved in water: 500 bars for a 5 N solution of caustic soda (NaOH) at 80°C and 560 bars for a 6 N solution of calcium chloride ($CaCl_2$) at 40°C.

For organic solutes in water, the data quoted in Landolt-Bornstein includes: 270 bars at 55.7°C for a solution containing 2.19 kg of cane sugar per liter, and 555 bars at 40.02°C for a 65% urea solution.

So far we have only considered osmotic pressures associated with aqueous solutions. The phenomenon is more general; there exists the certainty of osmotic pressure generation with any solution separated from its pure solvent by a semi-permeable membrane. Fiber reinforced resins are used to manufacture storage tanks for organic fluids, petroleum products for example, in which there exists the very real possibility of osmosis resulting from permeation by the fluid and dissolution of salts contained in the fiber reinforced resin. To illustrate the magnitudes of osmotic pressures associated with solutions of salts in organic solvents, we quote, again from Landolt-Bornstein (1923), (1936): 100 bars for the osmotic pressure of a 14.1% solution of nitrocellulose (gun cotton) in acetone, and 105 bars for a 24% solution of calcium chloride in alcohol.

Osmotic pressure can, of course, be calculated from the lowering of the vapor pressure of solutions by solutes for which there exists a vast amount of published data. The osmotic pressure is given by

$$p = \frac{RT\varrho_{solution}}{M_{solvent}} \ln \left[1 - \frac{\Delta p}{P_{solvent}} \right] \tag{10.6}$$

where $\varrho_{solution}$ is the density of the solution, and $M_{solvent}$ and $p_{solvent}$ respectively are the molecular weight and vapor pressure of the solvent. Δp is the measured lowering of the vapor pressure.

Lowering of vapor pressure data by solutes have been compiled for aqueous and non-aqueous solutions in Landolt-Bornstein (1905), (1923), (1927), (1936). The curves shown in Figure 10.9 are obtained by using Equation (10.6) to calculate p from Δp data quoted by Landolt-Bornstein for saturated solutions of anhydrous Na_2SO_4, of $Na_2SO_4 \cdot 7H_2O$ and of $Na_2SO_4 \cdot 10H_2O$. Since measurements for $\varrho_{(solution)}$ are known only for a few $Na_2SO_4 \cdot 10H_2O$ solutions at room tempera-

[45]Landolt-Bornstein. *Physikalisch-Chemische Tabellen* (1923), (1936).

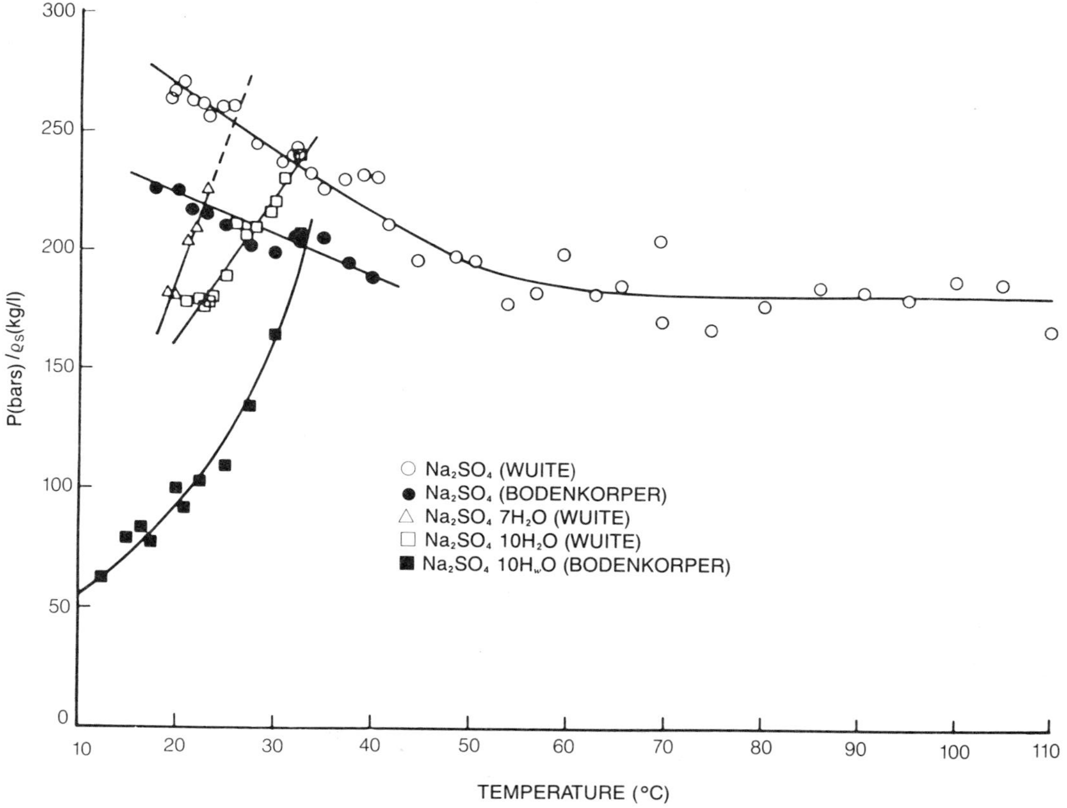

300
250
200
150
100
50
0
P(bars) / ϱs(kg/l)
10
20
30
40
50
60
70
80
90
100
110
TEMPERATURE (°C)
Na2SO4 (WUITE)
Na2SO4 (BODENKORPER)
Na2SO4 7H2O (WUITE)
Na2SO4 10H2O (WUITE)
Na2SO4 10HwO (BODENKORPER)

Figure 10.9. Osmotic pressures for solutions of Glauber's salt.

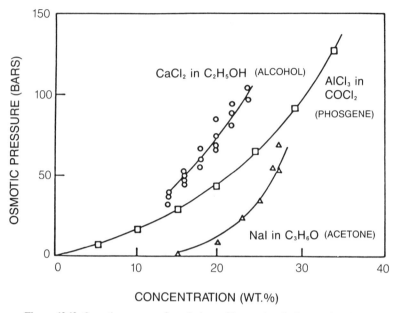

Figure 10.10. Osmotic pressures for solutions of inorganic salts in organic solvents.

ture, the data are presented as $P/\varrho_{solution}$. $\varrho_{solution} = 1.21$ kg l^{-1} at 20°C for a solution of Na$_2$SO$_4$·10H$_2$O containing 22% by weight of anhydrous salt, so the coordinate has to be scaled up by at least 20% for the numbers to represent osmotic pressure in bars. Further increases by some factor between 1 and 3 are required in order to take into account the ionic dissociation of Na$_2$SO$_4$ into three ions when in solution.

Figure 10.10 shows osmotic pressure data as a function of inorganic solute concentration for three organic solvents, namely calcium chloride in alcohol, aluminum tetrachloride in phosgene and sodium iodide in acetone.

10.5 Osmosis and loss of load transfer in fiber reinforced resins[46]

Substituting in Equation (8.42) the boundary conditions $P = 0$ at $z = 0$ and $z = l$, where l is the length of the fiber, it is found that the axial stress in an

[46]K. H. G. Ashbee and R. C. Wyatt. "Water Damage in Glass Fiber/Resin Composites," *Proc. Roy. Soc.*, A312:553–564 (1969).

embedded short fiber (see Figure 10.11) is

$$\sigma = \frac{P}{A_f} = E_f e \left\{ 1 - \frac{\cosh(\beta)\left(\frac{l}{2} - z\right)}{\cosh(\beta)} \right\} \tag{10.7}$$

This equation, representing the distribution along the fiber of axial fiber stress is sketched in Figure 10.11. Instead of Equation (8.38), the distribution of axial stress in the fiber can be described by

$$\frac{dP}{dz} = -2\pi r \tau \tag{10.8}$$

where τ is the interfacial shear stress, and r is the fiber radius.

Using $P = \pi r^2 \sigma$, H. L. Cox[31] shows that:

$$\tau = E_f e \sqrt{\frac{G_m}{E_f \cdot 2 \ln\left(\frac{R}{r}\right)} \cdot \frac{\sinh(\beta)\left(\frac{\ell}{2} - z\right)}{\cosh(\beta)\left(\frac{\ell}{2}\right)}} \tag{10.9}$$

where G_m is the shear modulus of the matrix material, and R is the radius of the composite cylinder.

Equation (10.9), representing the distribution along the fiber of interfacial shear stress, is also sketched in Figure 10.11.

Note in Figure 10.11 that $\sigma = 0$ at the fiber ends (as it must because there exists no fiber beyond its ends) and increases to a plateau in the middle region of the fiber. This gives rise to the concept of a minimum length of fiber below which fiber reinforcement is not fully utilized. Note also in Figure 10.11 that this simple model for fiber reinforcement predicts that the interfacial shear stress is extremely large at the fiber ends. In practice, yield or interfacial fracture occurs when τ at the fiber ends becomes large as can be seen in Figure 10.12 which shows the inward migration during environmental stress cracking, of flares of white photoelastic contrast defining the limits along a graphite fiber between which interfacial load transfer is effective. τ falls to zero at the points where σ reaches its maximum in long fibers.

The distributions of axial normal stress and of interfacial shear stress described by Equations (10.7) and (10.9), and shown graphically in Figure 10.11, have been dubbed the shear lag equations for fiber reinforcement.

Differential thermal contraction between fibers and resin during cooling from the resin cure temperature amounts to a net shrinkage of the resin on to the fibers;

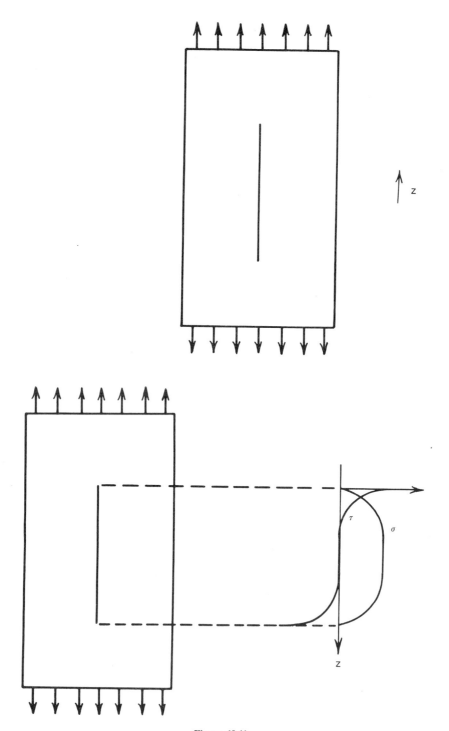

Figure 10.11.

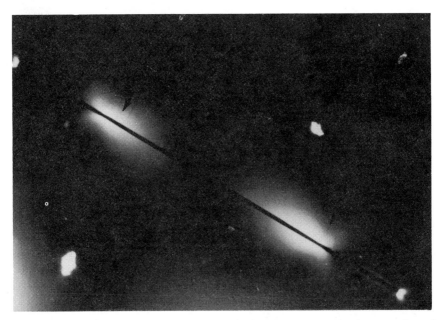

Figure 10.12. Graphite fiber in epoxy resin matrix after hot water exposure showing two regions of migrating resin birefringence (arrowed). Courtesy of E. Walter.

thus, the as-cast composite is self-loaded. The axial principal stress in each fiber is compressive and, for totally embedded fibers, is due partly to the load transferred across the fiber ends and partly to the shear stress transferred across the cylindrical part of its surface. The magnitude of the latter component increases from zero at the fiber ends to a maximum at its center. Now the optical retardation measured through a fiber diameter in a transparent glass fiber composite is due to the difference between the axial and tangential principal stresses, and since the latter is approximately constant with fiber length, the diametral optical retardation should increase from the fiber ends towards the fiber center, in line with the increase in axial compressive stress. After subtracting the retardation measured in the adjacent resin, this is found to be the case, Figure 10.13. The difference between the retardations through diameters at the fiber center and ends is a direct indication of load transfer. When loss of load transfer ensues, this difference approaches zero. Using this difference as a load transfer index, the onset and progress of debonding during water uptake can be followed in detail. Figure 10.14 shows data obtained for a polyester resin/E glass composite. That coupling agents alleviate this aspect of degradation by diffused water is demonstrated in Figure 10.15.

Loss of load transfer causes all of the resin shrinkage to be borne by normal stress transferred across the fiber ends and this is reflected by a large rise in retardation measured in the resin adjacent to the fiber ends, see Figure 10.14. This fact

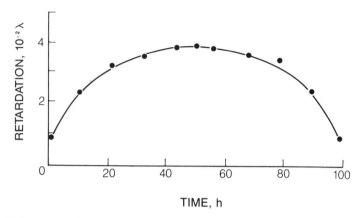

Figure 10.13. Variation of optical retardation measured through diameters along the length of a short fiber.

permits monitoring of loss of load transfer to opaque fibers, as shown in Figure 10.16 for a carbon fiber composite.

Loss of load transfer during water uptake is often initiated by the growth of discontinuous bubble-like regions at the fiber/resin interface, see Figure 10.17. The sign of birefringence adjacent to individual bubbles corresponds to that expected for pressure filled bubbles. E glass contains water-soluble constituents, and dissolution of this material gives rise to osmotic pressure of sufficient magnitude to cause debonding. Taking as trial value a continuous interfacial gap of 0.5 μm filled with aqueous solution, produced by leaching all available K_2O and Na_2O (combined content 0.6 wt%) from the fiber material, a 4 N caustic potash/caustic soda solution is expected, the associated osmotic pressure for which is large enough to grow resin cracks in hot wet resins. Moreover, the alkalinity of such a solution, besides chemically aiding fracture at the interface, enhances the dissolution of B_2O_3 from the fiber to form boric acid, hence boosting the osmotic pressure.

The occurrence of osmosis is inhibited by the presence of dissolved salts in the aqueous environment. Variations in the glass fiber composition also affect the incidence and rate of fiber debonding; sodium and potassium contents are of particular importance because of the high solubilities of their hydroxides and the strong alkalinity of the resultant solutions. In boiling water, total debonding of C glass fibers requires immersion times which are approximately one half of those for E glass fibers, although debonding occurs in an identical manner. This result is as expected from osmotic pressure considerations, since C glass contains about 9.6 wt% Na_2O compared with about 0.6 wt% $Na_2O + K_2O$ in E glass.

With fused pure silica fibers, debonding in the form of discontinuous bubbles is not observed during boiling water immersion. This is expected since pure silica contains no water soluble constituents.

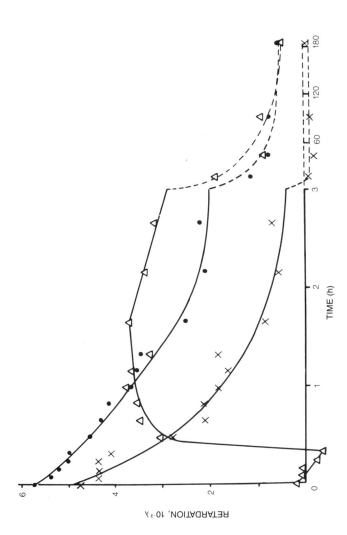

Figure 10.14. Optical retardation changes during immersion in water at 20°C. Untreated E glass fiber in polyester resin: × load transfer index, △ retardation in resin adjacent to fiber ends, • retardation through fiber center.

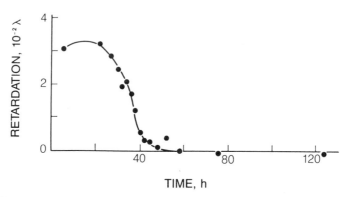

Figure 10.15. Long-term optical retardation changes arising from boiling water immersion. E glass fiber treated with Union Carbide A174 coupling agent.

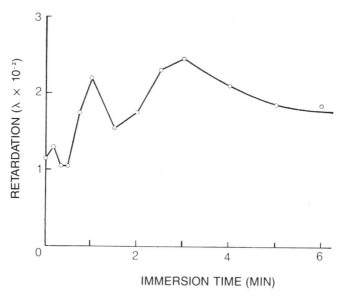

Figure 10.16. Optical retardation in the resin adjacent to the ends of a carbon fiber in a polyester resin composite exposed to boiling water.

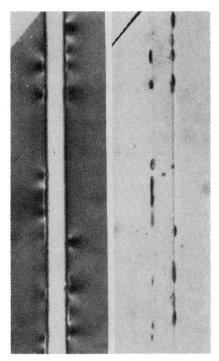

Figure 10.17. Transmission optical micrographs showing discontinuous debonding of an E glass fiber treated with a coupling agent: (a) ordinary light, (b) between crossed polaroids with compensator set to enhance the difference in birefringence.

Discontinuous debonding of glass fiber is often accompanied by the growth of cracks into the adjacent resin. In the case of E glass, these cracks are small and their growth ceases after about 20 hours of immersion in boiling water. For C glass fibers, the resin cracks nucleated at the interface are very much larger and more numerous and continue to grow over long periods of immersion, with the result that, in regions with a moderately high density of fibers, massive networks of intersecting cracks are formed. Examples of radial cracks of this kind are shown for both types of fiber in Figure 10.18. No radial cracks have been reported in composites made with fused silica fibers even after boiling water immersions of up to 500 hours.

10.6 Equilibrium shape and growth of osmotic pressure filled cracks

The equilibrium shape of a pressure-filled cavity in a rigid solid is flat and not spherical. F. C. Frank[47] estimates the pressure required to stabilize the equilibrium shape as follows.

[47]F. C. Frank. "The Equilibrium Shape of Gas-Filled Voids in a Solid," *J. Nucl. Mater.*, 48:199–200 (1973).

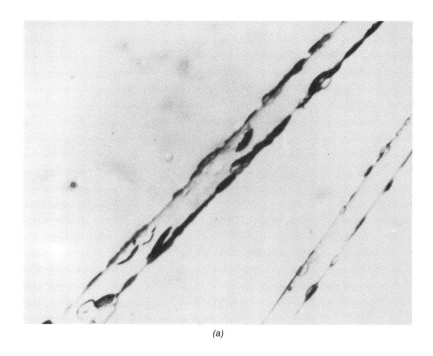

(a)

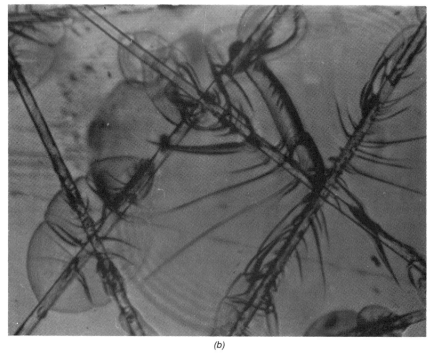

(b)

Figure 10.18. Polyester resin cracks accompanying debonding of fibers treated with a coupling agent. Specimens immersed in boiling water for 20 hours: (a) E glass, (b) C glass.

In an isotropic medium of Young's modulus E, and Poisson's ratio ν, the elastic increase in volume of a spherical cavity of radius a, due to an internal pressure p, is

$$\Delta V_{elastic} = 2\pi(1 + \nu)a^3 p/E \tag{10.10}$$

Since the distribution of stress in the neighborhood of a crack formed by a uniaxial tensile stress σ acting normal to its plane does not differ significantly from that given by a hydrostatic pressure $p = \sigma$ inside the crack (I. N. Sneddon[48]), the elastic increase in volume for a thin flat cavity may be calculated from the Griffith energy

$$U_G(\sigma) = \frac{1}{2} \sigma \Delta V_{elastic}(\sigma)$$

Thus

$$\frac{\Delta V_{elastic}(p)}{p} = \frac{\Delta V_{elastic}(\sigma)}{\sigma} = \frac{2U_G(\sigma)}{\sigma^2} \tag{10.11}$$

and the elastic stored energy is $\frac{1}{2} p\Delta V_{elastic}$. For a penny-shaped crack of radius a, R. A. Sack[49] found that

$$U_G(\sigma) = \frac{8(1 - \nu^2)a^3\sigma^2}{3E} \tag{10.12}$$

$$\Delta V_{elastic} = \frac{16(1 - \nu^2)a^3 p}{3E} \tag{10.13}$$

which is greater than $\Delta V_{elastic}$ for a spherical cavity and therefore indicates that the penny-shaped crack is the more stable configuration. The corresponding free energy decrease is

$$\frac{1}{2} p\Delta V_{elastic} = \frac{8(1 - \nu^2)a^3 p^2}{3E} \tag{10.14}$$

Hence the total free energy change due to formation of the thin flat cavity is

$$\Delta G = -pV_0 - \left\{ \frac{8(1 - \nu^2)a^3 p^2}{3E} \right\} + 2\pi a^2\gamma \tag{10.15}$$

[48]I. N. Sneddon. "The Distribution of Stress in the Neighborhood of a Crack in an Elastic Solid," *Proc. Roy. Soc.*, A187:229–260 (1946).
[49]R. A. Sack. "Extension of Griffith's Theory of Rupture to Three Dimensions," *Proc. Phys. Soc.*, 58:729–736 (1946).

where V_0 is the unstrained volume and γ is the specific surface free energy.

$$\frac{d(\Delta G)}{da} = \frac{-8(1 - v^2)a^2p^2}{E} + 4\pi a\gamma = 0 \qquad (10.16)$$

corresponding to a maximum G when

$$a = a^* = \frac{\pi E\gamma}{2(1 - v^2)p^2} \qquad (10.17)$$

for which value

$$\Delta G = \Delta G^* = -pV_0 + \frac{\pi^3 \gamma^3 E^2}{6(1 - v^2)^2 p^4} \qquad (10.18)$$

Taking as trial values, $E = 5$ GN m^{-2}, $\gamma = 200$ J m^{-2}, $v = 1/3$ for polyester resin, Equation (10.17) gives for the pressure required to nucleate a 2 mm diameter by 1 μm thick pressure-filled crack

$$p^2 = \frac{\pi \times 5 \times 10^9 (\text{N/m}^2) \times 200(\text{J/m}^2)}{2 \times 0.9 \times 10^{-3}(\text{m})}$$

$$= 1.7 \times 10^{15}(\text{N/m}^2)^2 \text{ since N} = \text{J/m}$$

That is

$$p = 4 \times 10^7(\text{N/m}^2)$$

$$= 400 \text{ bars}$$

which is of the order of magnitude estimated in Section 10.3 for osmotic pressure filled cracks.

The energy required to secure homogeneous transition from a compact (spherical) shape to the flattened form is found from Equation (10.18) to be $\cong 4 \times 10^{-4}$ J which is many orders of magnitude more than the energy available from thermal fluctuation ($kT \cong 4 \times 10^{-21}$ J at room temperature). Hence it is concluded that the flat cavities are not homogeneously nucleated. Instead, there must exist preferential orientations for cavity flattening. One possibility is local shear stress on the cavity plane.

The elastic nature of osmotic pressure filled cracks in epoxy and polyester resin and, in particular, the absence of any significant permanent (i.e., plastic) contribution to crack opening displacement, is easily demonstrated by photographing cracks and re-photographing them after drying. Drying causes the crack faces to come back into contact with each other, Figure 10.19 and 10.20.

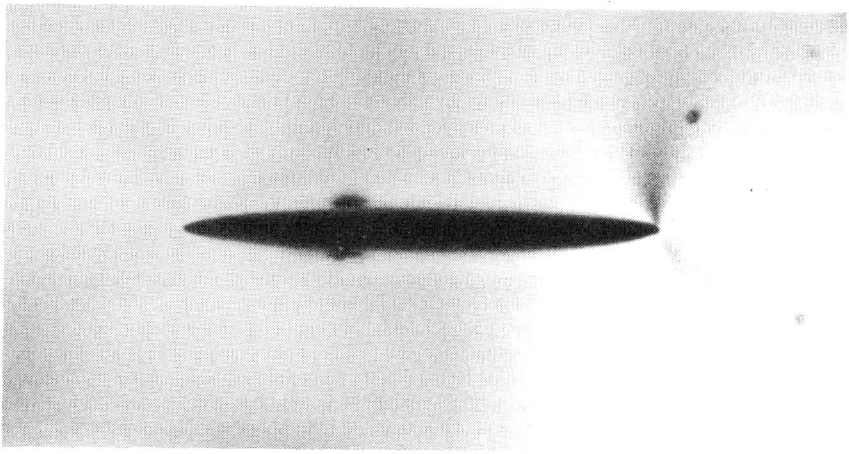

(a)

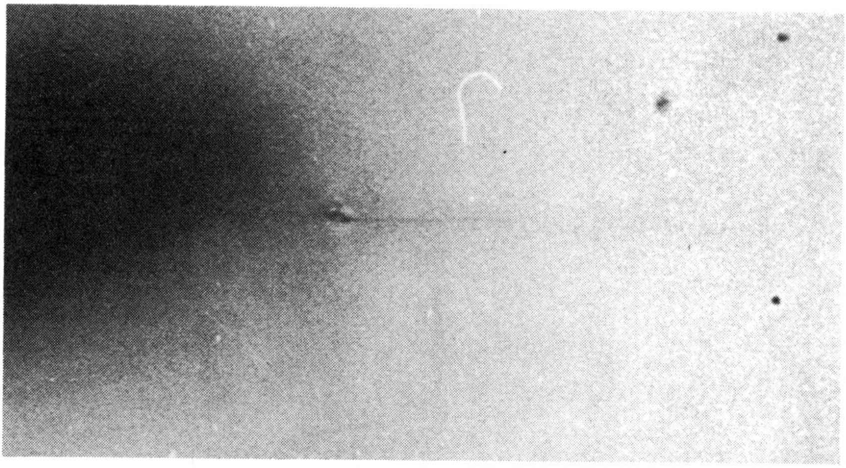

(b)

Figure 10.19. (a) Penny-shaped crack in KCl-doped epoxy resin photographed edge-on, 140 h immersion in water at 94 °C. (b) Same crack after drying in air for 1 h at 100 °C. (c) Face-on view after drying showing only three interference fringes, that is a permanent opening in the middle of the crack of only three half wavelengths.

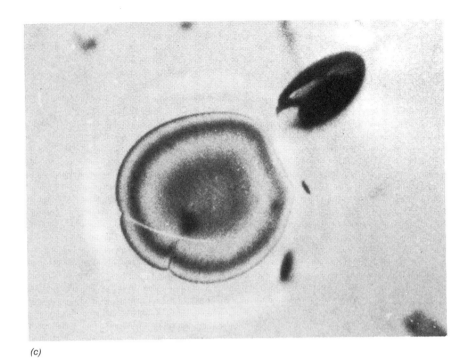

(c)

Figure 10.19. continued.

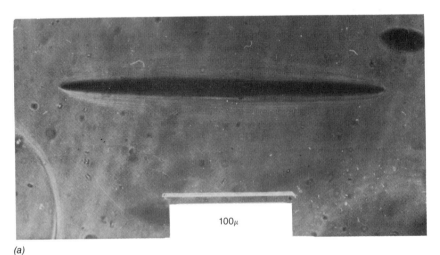

100μ

(a)

Figure 10.20. (a) Edge-on view of penny-shaped crack in polyester resin during hot water immersion. (b) Same crack after drying. (c) Face-on view after drying, showing interference fringes.

294

(b)

(c)

Figure 10.20. continued.

Seen face-on, such cracks in dried samples generate interference patterns and, by counting the fringes, the very close proximity of the crack faces can be measured; the small residual space between the crack faces is attributed to obstruction by solute deposited during drying.

Figure 10.21 shows edge-on views of a penny-shaped crack in epoxy (a) and polyester resin (b) samples photographed after several times of immersion in water at 94°C. Measurements of the crack diameter reveal[50] that the radial crack growth rates in epoxy resin at 70°C and 80°C are constant at 4.9×10^{-10} m s^{-1} and 1.1×10^{-9} m s^{-1}, respectively. In polyester immersed in water at 70°C and 80°C constant crack growth rates of 5×10^{-12} m s^{-1} and 7×10^{-11} m s^{-1}, respectively have been reported.

The stress intensity factors deduced from the geometries of the cracks shown in Figures 10.21a and 10.21b lie between 0.23 and 0.79 MN m$^{-3/2}$.

10.7 Water of hydration at impurity inclusions

". . .when an anhydrous salt unites with a definite number of water molecules to form the crystalline hydrate . . . the water molecules change their state of liquid aggregation to become the constituents of a solid body. . ."[51]

Phenomena by which diffused water can generate pressure at impurity inclusions include hydration and gas evolution as well as osmosis. Here we examine the phenomenon of hydration in polymer matrix composites.

Many common fillers can exist in thermodynamic equilibrium in more than one state of hydration. By virtue of the space occupied by water of crystallization, hydration from a lower to a higher hydration state is always accompanied by a volume increase (ΔV) which, for a particle of filler encapsulated by polymer, generates a confining pressure (p) from the polymer matrix. Since hydration immobilizes molecules of liquid water into a crystal lattice, there is also an associated decrease of entropy, known as the entropy of hydration (ΔS). The equation

$$\frac{dP}{dT} = \frac{\Delta S}{\Delta V} \tag{10.19}$$

derived by Clapeyron, applies to any change in which a volume increase working against a confining pressure is accompanied by an entropy change, and states the dependence of the equilibrium value of this pressure on temperature. Given its temperature coefficient, we can calculate the equilibrium pressure at any temperature if we know it at one temperature. Here, we make use of the phase diagram

[50]J. P. Sargent and K. H. G. Ashbee. "Very Slow Crack Growth During Osmosis in Epoxy and Polyester Resins," *J. Appl. Polymer Sci.*, 29:809–822 (1984).
[51]J. Thomsen. *Thermochemische Untersuchungen III* (1883), *Termoken Resultaker* (1905), English translation by K. A. Burke, Longmans, Green, London (1920).

68 HRS

92 HRS

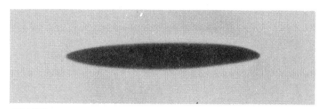

116 HRS

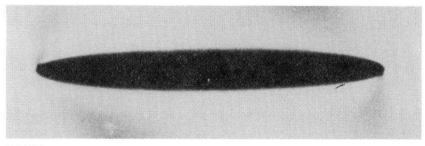

139 HRS

|—————————————————————————————| 100μm

Figure 10.21a. Edge-on view of growth of single penny-shaped crack ~ 0.2 mm below the surface in epoxy resin.

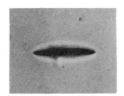

45 HRS

51 HRS

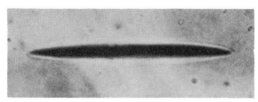

68 HRS

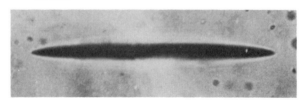

76 HRS

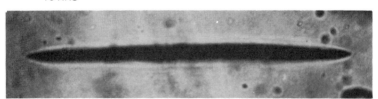

90 HRS

97 HRS

|_____| 100μm

Figure 10.21b. Edge-on view of growth of single penny-shaped crack ~ 0.2 mm below the surface in polyester resin.

298

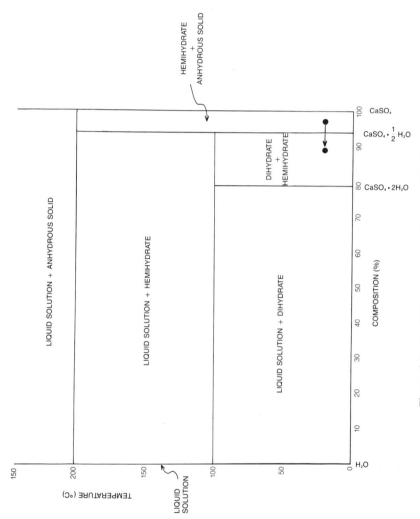

Figure 10.22. Equilibrium diagram for the system H_2O–$CaSO_4$.

299

for a common filler $CaSO_4 - H_2O$ system, shown in Figure 10.22, to find a baseline pressure from which to estimate an order of magnitude for a typical water of crystallization pressure.

There is a range of temperatures, 100°C downward, in which dihydrate is in equilibrium with liquid solution. There is one temperature, 100°C, at which dihydrate and hemihydrate together are in equilibrium with liquid solution. Above 100°C, liquid solution is in equilibrium with hemihydrate (and the solubility, not shown, diminishes with increase of temperature). Anhydrous solid $CaSO_4$ can co-exist in equilibrium with liquid solution above 200°C. At any given temperature, however, there is a partial pressure of water vapor in equilibrium with co-existing anhydrous $CaSO_4$ and hemihydrate. At temperatures below 200°C this partial pressure is less than that in equilibrium with any aqueous solution of $CaSO_4$.

J. Thomsen[51] measured the heats of hydration of many common inorganic salts. Other measured values were compiled and published in Landolt-Bornstein early in this century. Taking average published measurements, per water molecule, we have for the dihydrate at room temperature

$$\Delta H = 4.89 \text{ kcal per mole}$$

Hence

$$\Delta S = -\frac{\Delta H}{T} = -\frac{4.89 \times 10^3 \text{ (cal)}}{293 \text{ (K)}} = -16.70 \times 4.18 \text{ (J/K)}$$

Taking the density of $CaSO_4 \cdot 2H_2O$ to be 2,320 kg m^{-2}, the volume per water molecule in 1 gm mole is $\Delta V = 14.09$ cm^3. This compares with 18 cm^3 for the natural volume of 1 gm mole of free water.

Substituting for ΔS and ΔV in Equation (10.19) we find

$$\frac{dP}{dT} = \frac{-16.70 \times 4.18 \text{ (J/K)}}{14.09 (\text{cm}^3)}$$

$$= -4.94 \text{ (J/K cm}^3)$$

$$= -4.95 \times 10^6 (\text{J/K m}^3)$$

$$= -49.5 \text{ (bar/K)}$$

Negative dP/dT means that the pressure increases with decrease of temperature.

From Figure 10.22 it is evident that at 100°C the hemihydrate, the dihydrate and a liquid solution containing 0.01% by weight of calcium sulfate would be in equilibrium. At this temperature, therefore, the equilibrium pressure bearing on an inclusion consisting of hemihydrate and dihydrate would be zero if the external aqueous environment were that dilute liquid solution. When the external aqueous environment is pure water, a source of water molecules at a higher chemical potential, the equilibrium pressure is higher, by an amount equal to the osmotic pressure of the 0.01% solution (of order 10 bar, say). To this must be added a further 495 bar for every 10° drop in temperature below 100°C.

Above 100°C the calculation involving the hemihydrate does not apply. Hemihydrate may be assumed to have formed already, and the hydration product at the inclusion is liquid solution in equilibrium with crystalline hemihydrate; the equilibrium pressure corresponds to the osmotic pressure of this solution.

On cooling below 100°C* the liquid inclusion can solidify to dihydrate (also consuming a little more of the hemihydrate). Assuming that some hemihydrate still remains, the inclusion can continue to grow by receipt of more water, converting to dihydrate and develop enhanced pressure as the temperature falls.

Of course it is not certain, or even likely, that dihydrate appears as soon as it becomes the stable phase. Liquid solution is likely to persist in a supercooled state through some finite temperature range, and until the dihydrate is nucleated the additional pressure due to its formation will not be generated.

A typical calcium sulfate filler might have the composition 50% anhydrous solid plus 50% hemihydrate, and here it is instructive to consider the implications of plaster of Paris setting to gypsum. In Figure 10.22 this 50/50 composition is indicated by a point at ambient temperature (20°C) in the middle of the hemihydrate plus anhydrous solid, two-phase field. Suppose that after prolonged use in a humid environment, there is sufficient water uptake by the polymer matrix to move the local composition to a point within the dihydrate plus hemihydrate, two phase field (refer to Figure 10.22). 20°C is 80°C below the temperature at which the water of crystallization pressure would be zero for this composition. So, if sufficient time has elapsed for hydration to the dihydrate plus hemihydrate phase mixture that corresponds to the new local composition, an internal pressure, localized at each particle of filler, of some 2.5 kbar would have been generated. This is far in excess of the tensile strength, even of dry polymers at 20°C, so internal cracking would certainly have occurred.

In passing, it should be noted that reactions similar to hydration can take place with substances other than water. Alcoholation and ammoniation are two examples. Since there is essentially no difference between such reactions and hydration, pressure generation akin to that which accompanies formation of water of crystallization is expected.

Hydratable salts have also been identified in the surface layers of PAN graphite fibers (the PAN precursor is spun from sodium thiocyanate solution and traces of sodium carried through to the carbonization stage are converted to sodium carbonate-soda), in aramid fibers (sodium sulfate—Glauber's salt) and in adhe-

---

*A source of water at higher chemical potential than in the saturated calcium sulfate solution (e.g. pure water) would favor the more hydrated state. The temperature below which the dihydrate is stable with respect to the hemihydrate would then be somewhat higher than 100°C. This higher temperature is the temperature relative to which ΔT should be calculated.

sive joints involving metals which have been anodized in phosphoric acid (disodium hydrogen phosphate).

The respective temperature coefficients of pressure associated with hydration to higher hydrates are:

dP/dT = -20.5 bar K^{-1} per water molecule for $Na_2CO_3 \cdot 7H_2O \rightarrow Na_2CO_3 \cdot 10H_2O$ in graphite fiber composites,

dP/dT = -11.3 bar K^{-1} per water molecule for $Na_2SO_4 \rightarrow Na_2SO_4 \cdot 10H_2O$ in aramid fiber composites,

dP/dT = -19 bar K^{-1} per water molecule for $Na_2HPO_4 \rightarrow Na_2HPO_4 \cdot 12H_2O$ in adhesive joints containing phosphate deposits.

10.8 Worked examples

1. The fact that pockets of free water, and of solutions of free water, can exist within resin-based composites, combined with the fact that a volume expansion of some 8.4% accompanies the freezing of water, raises the possibility of an internal fracture process akin to frost heaving of stone aggregates. Estimate the magnitude of the internal pressure generated during freezing. The entropy of fusion of ice is $\Delta S = 1.225$ J $gm^{-1}K^{-1}$.

Like hydration, freezing immobilizes molecules of liquid water into a crystal lattice so there is an associated decrease of entropy and, since the volume expansion works against a confining pressure, Clapeyron's equation, Equation (10.19), can be used to estimate the magnitude of the pressure generated.

The density of ice is 0.92 gm cm^{-3} so the change in volume that occurs when 1 gm of water freezes is $\Delta V = 1 - 1/0.92 = 0.09$ cm^3.

Thus

$$\frac{dP}{dT} = \frac{\Delta S}{\Delta V} = \frac{-1.225 \left(\dfrac{J}{gm\ K} \right)}{0.09 \left(\dfrac{cm^3}{gm} \right)}$$

$$= -13.61 \left(\frac{bar}{K} \right)$$

It is evident that a fall in temperature to, say, $20°$ below the freezing point would create internal pressures of $\cong 272$ bar.

2. Explain why occluded water in fiber reinforced plastics that have been exposed to aqueous environments, does not boil and hence initiate shattering during the thermal spike experienced by aircraft subjected to supersonic dash.

Almost certainly, aqueous occlusions in fiber reinforced plastics are of aqueous solutions and not pure water. Osmotic pressures for aqueous solutions of inorganic salts can be very high, and it is very likely that the internal pressures associated with aqueous solutions will exceed the critical pressure. If true, there is no possibility that the water can boil or even evaporate. The reasoning for this goes as follows.

Depending on the pressure, water at any temperature below the critical point may exist in the gaseous state (low pressures) or in the liquid state (high pressure). At some intermediate pressure, liquid and vapor may be present simultaneously. When both phases co-exist, the saturated vapor is in thermodynamic equilibrium with the liquid. On a p, v diagram an isothermal for such a system will have the form shown in Figure 10.23.

Region A'D (for which the pressures are low) corresponds to vapor alone. The horizontal section AA' corresponds to a mixture of saturated vapor and liquid in varying proportions given by the "see-saw" rule. Region AB (over which the pressures are high) corresponds to liquid alone. Since liquids are only slightly compressible, region AB rises steeply as v decreases.

The equation of state can be represented on the p, v diagram by a family of isothermals whose general aspect, including transition to the liquid state and the region around the so-called critical point, is illustrated in Figure 10.24.

At temperatures below that of the critical point C, the isothermals have the form already discussed in Figure 10.23. At somewhat higher temperatures the horizontal section, corresponding to a mixture of liquid and saturated vapor, has become shortened from AA' to A_1A_1'. At the critical point, the horizontal section has contracted into a point of inflexion at which the tangent is horizontal.

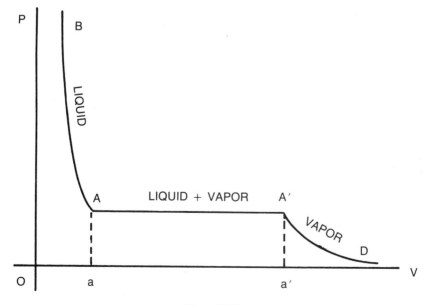

Figure 10.23.

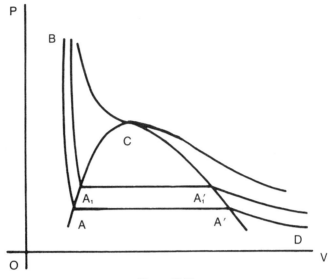

Figure 10.24.

The critical points for water are: critical temperature 374°C, critical pressure 218 atmospheres, critical volume 3.2 cm³/gm.

If the osmotic pressures associated with internal pockets of aqueous solutions exceeds 218 atmospheres then, however high the thermal spike temperature, there is no possibility that the solution can boil. Raising the temperature will, of course, increase the magnitude of the osmotic pressure, but only in accordance with the gas laws.

10.9 Coupling agents

In view of the need for efficient and durable load transfer between matrix and fibers, a great deal of attention has been paid to the molecular nature of the interfacial region in fiber reinforced plastics, which although too small in extent to be recognized as a phase in the Willard Gibbs' sense of the word (a phase is something to which the Gibbs phase rule can be applied), is nevertheless often referred to as the interphase. Resins shrink during curing and radially compress the fibers, and this compression is enhanced by the differential thermal contraction between fiber and matrix materials during cooling from the resin cure temperature. To some extent this radial compression is relieved by resin flow although, since some fibers (graphite) are better heat sinks than others (glass), the interfacial temperature and hence the interfacial rheological behavior during exothermic cure reactions is different for different composites.

The formation of chemical bonds is characterized by factors ordinarily associated with chemical reactions, that is by activation energies leading to reactions with finite rates, irreversibility, and heats of reaction large compared with those for physical interactions. On a molecular scale, the formation of a chemical bond is accompanied by large changes in electronic structure and often by electron transfer, whereas physical interactions are characterized by much weaker electronic perturbations. Fiber/resin systems which display particularly good interfacial adhesion are glass fiber/polyester resin, asbestos fiber/phenolic resin, and carbon fiber/epoxy resin.

With a view to promoting chemical bonding between matrix and fiber materials, fibers are usually coated with organofunctional silanes.[52] These coupling agents contain functional groups that can react chemically with silanol groups on glass fiber surfaces, and other functional groups which, in principle, could react chemically with, for example, epoxy groups in the matrix. In this way, a measure of chemical bonding is achieved between fibers and matrix. The results are often spectacular, especially in respect to durability of interfacial load transfer in composites exposed to aqueous environments (see Figure 10.15). In fact, so successful are the proprietary coupling agents that their application has been extended to the adhesive bonding to resins of oxide surfaces other than glass, e.g., oxide films grown on metals and even thermally etched (oxidized) graphite fiber surfaces.

10.10 Indentation cracks following debonding in short fiber composites

Debonding permits relative longitudinal movement between fiber and resin and, if the overall axial stress is compressive, fiber ends behave as rigid indenters and propagate conically shaped cracks into the adjacent resin.

Figure 10.25 shows this phenomenon in an E glass/polyester resin composite in which resin cracking at the ends of the fibers continues to develop during the resin shrinkage that accompanies prolonged exposure to hot water. These cracks are not unlike Hertzian cracks (see Chapter 8, Section 8.6). After the nucleation and growth of a first crack, further water uptake causes the polyester resin to continue to shrink and, at each crack, the fiber end penetrates the near face of the crack and then pushes against and displaces the outer face. Crack widening thus occurs and continues until the outer surface is eventually pierced. The stress field in the resin beyond the first crack again builds up, a second crack is nucleated, and the crack widening, displacing and piercing process repeats itself. In this way, several cracks can be created in the resin at a fiber end. The process incurs large deformation and hence a high degree of birefringence in the resin adjacent to the fiber end, see Figure 10.26.

When short fiber polyester resin composites are exposed to water for long times and then fractured, the overall fracture surface includes indentation cracks, which can then be examined by scanning electron microscopy. Figure 10.27 shows an E glass fiber protruding through the inner surface of an indentation crack.

[52]See, for example, E. P. Pluddeman. *Silane Coupling Agents*, Plenum Press, New York (1982).

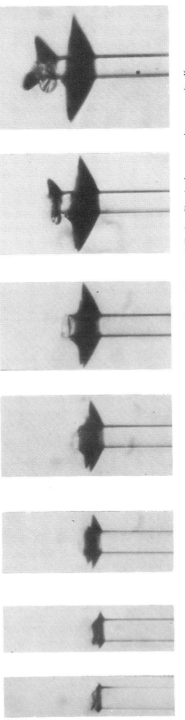

Figure 10.25. Growth of indentation cracks following debonding of short fibers. Untreated E glassfiber/polyester resin composite.[46]

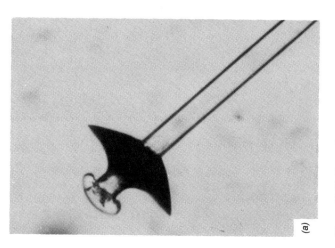

Figure 10.26. Transmission optical micrographs of indentation cracking: (a) ordinary light, (b) between crossed polaroids showing high birefringence in adjacent resin due to extensive plastic deformation.

Figure 10.27. Scanning electron micrograph of a 10μm diameter fiber protruding through the inner surface of an indentation crack. Untreated E glass fiber/polyester resin composite exposed to boiling water for 400 hours.[46]

None of the above cracks is due to application of external stress. They are simply the consequence of the self-stressing brought about by resin dimensional changes that accompany exposure to aqueous environments. The build-up of the stress field that leads to indentation cracks occurs more quickly the longer the fiber.

10.11 Further examples

1. Using examples from composite materials, describe each of the migration processes known as diffusion, intercalation, and reptation.
2. A square picture frame with mitred corners has wooden sides of length l and width b (measured in the plane of the frame). A change of humidity causes the width of the sides to increase by 1%, but causes no change in the length of the

sides; as a result, the frame distorts without breaking. Sketch the resulting shape and calculate the curvature of the side (University of Bristol Examination, 1970).

3. (a) A stressed, and therefore optically anisotropic, transparent fiber, embedded in resin, appears dark when examined in transmission between crossed polars. What factors would give rise to this effect and how would you distinguish between them? (b) Under what circumstances and how can you use a quarter wave plate to measure optical retardation? (c) In a case where the method under (b) is applicable, it is found that the analyzer must be rotated 22° from the crossed polars position in order to achieve extinction. What is the birefringence in the fiber if its thickness is 10^{-3} mm and light of wavelength 589 nm is being used (University of Bristol Examination, 1973)?

4. (Laboratory exercise) Prepare about 25 g of transparent (epoxy or polyester) resin mix and add to it 0.5 wt% of finely ground potassium chloride (KCl). Cast your resin mix into slabs with approximate dimensions 50 mm × 50 mm × 5 mm thick, and cure in accordance with the manufacturer's recommendations. Immerse the specimen slabs so obtained in boiling water for 1, 2 and 3 days, and describe any changes visible by eye. Examine any microstructural defects in a polarizing microscope, and describe your observations.

5. Environmental degradation of polymer matrix composites is less severe in saline than in fresh water environments. The opposite is true for both metals and metal matrix composites. Account for both observations.

6. Starting from first principles, argue the case that in thermodynamic equilibrium, alkali metal impurities in the surface layers of PAN-based carbon fiber must be present as carbonate rather than oxide or elemental form.

7. Osmotic pressures associated with aqueous solutions are usually calculated from measurements of the lowering by solutes of the vapor pressure above solutions. Derive from first principles the relationship between osmotic pressure and the lowering of vapor pressure, and use it to calculate the osmotic pressure for a saturated solution of sodium carbonate at 43°C given that its vapor pressure is 66 millibars.

8. Di-sodium hydrogen phosphate dihydrate ($Na_2HPO_4 \cdot 2H_2O$) is a product of neutralization of traces of sodium hydroxide during phosphoric anodizing of aluminum matrix composites to be joined by adhesive bonding. Sketch the $Na_2HPO_4 - H_2O$ equilibrium diagram given that the peritectic temperatures are as follows:

liquid solution + anhydrous solid → dihydrate   95.0°C
liquid solution + dihydrate → heptahydrate   48.0°C
liquid solution + heptahydrate → dodecahydrate   35.5°C

Hence, or otherwise, estimate an order of magnitude value for the pressure generated if water diffused through the adhesive converts dihydrate inclusions to dodecahydrate. The heat of hydration at 18°C is 119 kJ/mole.

9. It has been proposed that microwave radiation could be used to detect the

presence of "free" water in carbon fiber reinforced plastics. Taking the resistivity of carbon fiber as $10^{-3}\Omega$ cm and assuming that plastic is essentially lossless ($\epsilon \sim 3$) show that (i) the wavelength in the resin is very much larger than both of the fiber diameter and the surface to surface separation of fibers; (ii) the skin depth in the fibers is much larger than the fiber radius; (iii) with the electric field parallel to the fibers so that the effective dielectric constant is given by the law of mixtures —

$$\epsilon_{composite} = \eta\epsilon_{fiber} + (1 - \eta)\epsilon_{resin}$$

—where η is the fiber volume fraction, the attenuation length, that is the distance in which the wave amplitude is reduced to $1/e$ of its initial value, is of the order of a few tens of μm; (iv) with the electric field perpendicular to the fiber direction, the effective dielectric constant of the composite is

$$\epsilon_{composite} = \epsilon_{resin} - \frac{\eta}{\epsilon_{fiber}}$$

and the attenuation length is of the order of a km, so quite small losses in ϵ_{resin} due to water, particularly if concentrated at the fiber/resin interface since here the field is strongest, should be detectable with this polarization. (Note the inference from (iii) and (iv) that unidirectional carbon fiber reinforced plastic behaves as an efficient polarizer of microwave radiation.)

11

Joining and repair

*"...time and place often give the weak advantage
over the strong..."[53]*

The joints between the various component parts of structures are notoriously inefficient, in terms of both stiffness and strength. The best knots and splices in rope are only 40% to 80% as strong as rope itself. Nailed, screwed, pegged, dovetailed, and mortised and tenoned joints in wood are even less efficient. No more efficient are bolted and riveted joints in fiber reinforced composites. This is especially true for thin adherends; thin walled joints are notoriously prone to tearing at stress concentrations around mechanical fasteners. Therefore, it is not surprising that, where possible, mechanical connections of all kinds have to a large extent been displaced by modern adhesives.

11.1 Evolution of designs for joints

The evolution of designs for joints was central to the wood technologies of ancient societies. Many of the world's maritime museums possess examples of joints in archaeological remains of wooden boats. Somewhat more accessible are the fourteenth and fifteenth century scarfing joints in timbers employed in Europe in the construction of buildings still standing today. The geometries evolved for joint design in wood are empirical. They take into account non-parallelism of grain orientation between the members joined; for this reason, they are a good working basis from which to develop designs for joints involving fiber reinforced composite materials.

Figure 11.1 shows how timbers are halved and notched into one another. Mechanically, this is a weak joint because of the reduction, by one half, made in the thickness of both members and, more importantly, because of the stress concentrations of the re-entrant corners. Figure 11.2 shows a scarfed joint. In carpentry, various kinds of scarfed joints are used when it is necessary either to join the ends of timbers to increase their length or to join timbers to other materials, metals for example. An ordinary splayed scarf joint is made by cutting

---

[53]*Aesop's Fables*, 620 B.C.

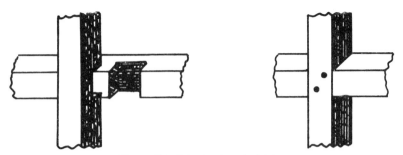

Figure 11.1. Halved and notched joint.

the two ends of the timbers to a long angle (slope 1 in 10), and glueing, dowelling, bolting or otherwise securing the two parts together. It is essential to have both angles alike and both faces square to each other. This joint is useful in general repair work. Halving, when adopted as a method of joining timbers longitudinally or end-to-end, is also described as scarfing. Half of each of the two ends to be joined is cut away longitudinally so that the two pieces can lap on to each other, see Figure 11.3. The length to which each end should be reduced has been proven by experience to be five to six times its depth, and about twice its breadth. This is a useful all-round joint and is easy to make. When used in timber, the lapped scarfed joint is usually secured with nuts and bolts. The iron fish plates shown in the sketch, though not always necessary, greatly increase the strength of the joint. Such a joint would not be suitable for an upright post which has to support a moving load.

The splayed scarfing joint, used for lengthening tie beams, is sketched in Figure 11.4. The ends are held in position, prevented by the projecting tongues from springing apart. Recommended proportions for the length of the joint are about three and a half times the greatest breadth. The tabled scarfing joint, Figure 11.5, has the advantage that it naturally resists longitudinal tension as well as longitudinal compression. Also, shrinkage of the timber does not seriously affect its security. However, it requires careful fitting; the sides of the timber have to be very carefully marked out and the lines accurately square and true.

In the ordinary splayed scarfed joint, Figure 11.4, the junction forms a straight line. In the modes of scarfing in Figure 11.5 it does not. The extremities of the upper timber in Figure 11.5(a) are in the line ACDB and of the lower timber in the line BEFA. In Figure 11.5(b), grooves of exactly the same size are cut in the faces of the timbers as at CD. Their locations are such that, when brought

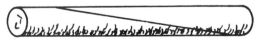

Figure 11.2. Scarfed joint.

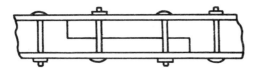

Figure 11.3. Lapped scarf joint.

Figure 11.4. Splayed scarf joint.

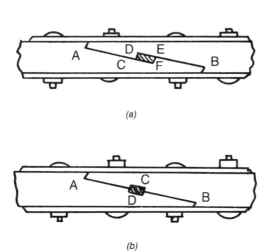

(a)

(b)

Figure 11.5. Tabled scarfing joints.

together, they nearly, but not quite, coincide. A taper wedge is driven into the double groove thus formed, and this has the effect of locking the shoulders AB into the angles cut for receiving them. In Figure 11.5(a) such a wedge bears against surfaces CD and EF and draws the ends of the pieces together, again forcing the extremities A and B into the angles cut to receive them.

11.2 Mechanical fastening

In principle, the ladders used by the Fire Department could be increased in length by 50% without any increase in weight, if made from graphite fiber reinforced epoxy resin instead of from aluminum alloy. However, much of the stiffness advantage would be lost by the extra compliance introduced at the joints to the extra sections of ladder.

Mechanical fasteners, that is rivets, screws and bolts, are used extensively to join composite structures in spite of the stress concentration around the holes in which fasteners are located (see Chapter 4). In all cases it is better to err on the side of interference fit than on the side of clearance between fasteners and holes.

Rivets are permanent fasteners. To spread the load and avoid pull-through, 100° or 120° countersunk heads are preferable to the conventional 90° heads used to fabricate thin-walled metal structures. Even better load spreading is obtained with mushroom heads. Doublers, of ±45° crossply laminates for example, are commonly used in the region of a joint.

Self-tapping screws are used in permanent joints and in joints which can be taken apart. They cut a tight thread especially in carbon fiber composites.

Bolts are the traditional means for de-mountable fastening. One useful rule of thumb with bolted joints is to make the diameter of the bolt greater than the thickness of the laminate even if this necessitates use of hollow bolts in order not to impose a weight penalty. The mechanical contact in a tightly bolted joint is superior to that in either riveted or screwed joints.

The advantages of mechanical joints over adhesively bonded joints include the capability for repeated assembly mentioned above (albeit at the expense of the loss in strength due to introduction of a hole to accommodate a bolt, rivet or screw), absence of environmental effects peculiar to polymers (such as inhomogeneous swelling in aqueous environments), and ease of inspection. In the case of mechanical fastening to laminates, there is the additional advantage of the stitching role inhibiting delamination.

Localized delaminations propagate from the edges of holes and cut-outs during fatigue loading, and present free surface to transverse cracks growing under the action of in-plane static loads. Since a crack has to stop when it meets a free surface, the consequence is a measurable increase of in-plane static strength. Table 11.1 shows data for [0/±45] boron fiber/epoxy resin laminates. The reduction in strength for the specimen that contained no holes is attributed to accumulation of fiber breakages during fatigue loading.

In carbon fiber reinforced plastics, a bolted joint establishes good electrical contact between bolt and fibers. In the presence of electrolytes an elec-

Table 11.1. (after J. C. Halpin, loc. cit.).

	Average static tensile strength before fatigue testing (MN m^{-2})	Average residual tensile strength after 5 × 10^6 cycles (MN m^{-2})
Control sample (no hole)	575	530
hole 1.5 mm diameter	475	535
hole 1.5 mm diameter with notch 0.5 mm depth × 100 m width	500	570
hole 1.5 mm diameter with notch 2 mm depth × 100 m width	400	540

trochemical cell is set up and, if the bolt is made from a light alloy, corrosion ensues. Stainless steel or better still, since it incurs no weight penalty, titanium bolts are preferred. This problem does not often occur with rivets or screws especially if they are cadmium or zinc dipped.

11.3 Galvanic corrosion of mechanical fasteners in carbon fiber composite structures

When different metals are joined, including graphite and silicon carbide as honorary metals, and then exposed to an electrolyte, galvanic corrosion can occur. Figure 11.6 shows, as example, the galvanic cell set up between iron and carbon fiber reinforced plastic in a saline environment. Iron passes into solution

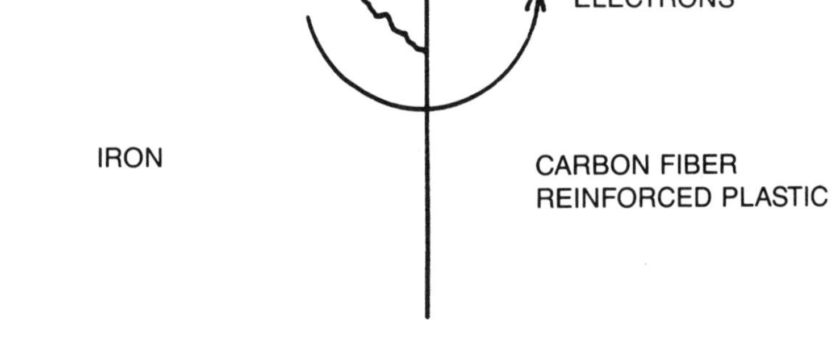

Figure 11.6. Galvanic corrosion between iron and carbon fiber reinforced plastic.

as Fe^{2+} ions, viz.,

$$\text{\textcircled{Fe}} \rightarrow Fe^{2+}$$
$$\searrow$$
$$2e$$

and the electrons left behind are conducted through and to the surface of the carbon fiber reinforced plastic, where the reaction

$$O_2 + 2H_2O + 4e \rightarrow 4OH^-$$

takes place, releasing OH^- radicals into the electrolyte. The Fe^{2+} ions and OH^- radicals combine to form ferrous hydroxide $Fe(OH)_2$ and, depending on the state of aeration in the solution, this may then oxidize to $Fe(OH)_3$, to $Fe_2O_3 \cdot H_2O$ (red-brown rust) or to Fe_2O_3 (black magnetite). In its mildest form the consequence is unsightly stains but, if serious, it is very damaging because it concentrates sharply on the less noble metal—the iron in this particular case—and produces localized, deep attack. At the junction, a large corrosion current is sustained because the electrical resistance of the short path length through the electrolyte is low.

It is not always possible to predict from the electrochemical table which of two metals when joined will be the anode and which the cathode. For example, if a highly reactive metal is covered by a protective oxide film, its ions cannot easily traverse it and go into solution even though the metal is readily ionizable. For this reason, titanium behaves like a noble metal in a galvanic cell, despite its position in the electrode potential scale, Table 11.2.

Half-cell potentials are determined against a hydrogen electrode which comprises a platinum electrode immersed in a 1 N solution of hydrochloric acid saturated with hydrogen gas at one atmosphere pressure. The platinum electrode takes up the potential at which dissociation into nascent hydrogen ($H_2 \rightarrow 2H$) is balanced by recombination of nascent hydrogen ($2H \rightarrow H_2$). The electrode potential of any metal (for example silver) is determined by placing it in a 1 N solute that contains its ions (for example $AgNO_3$), joining the half-cell so formed to the hydrogen electrode half-cell via a connection solution chosen to suppress liquid junction potentials, and measuring the potential that balances the emf of the combined half-cells. Figure 11.7 shows the set-up.

Stainless steel usually behaves as a noble, passive metal, but, if its oxide film is destroyed, it becomes reactive and much more susceptible to galvanic attack. By taking pairs of metals, including carbon, in a given environment such as salt water, and observing which becomes the anode, so-called galvanic series have been compiled. One such series for sea water corrosion is given in Table 11.3.

One advantage of adhesively bonded joints over mechanical joints is that the

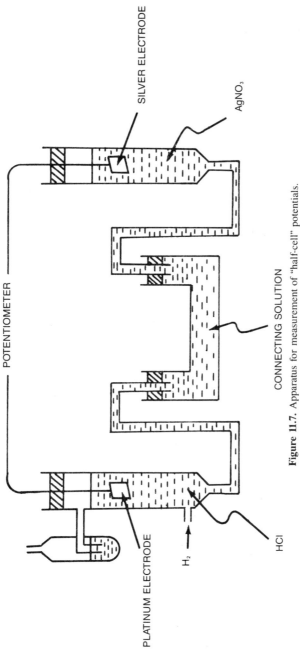

Figure 11.7. Apparatus for measurement of "half-cell" potentials.

Table 11.2. Standard electrode potentials, hydrogen scale, 25°C.

Metal	Ion	Volts
Cs	Cs^{1+}	− 3.02
Li	Li^{1+}	− 3.02
Rb	Rb^{1+}	− 2.99
K	K^{1+}	− 2.922
Na	Na^{1+}	− 2.712
Ca	Ca^{2+}	− 2.5
Mg	Mg^{2+}	− 2.34
Al	Al^{3+}	− 1.67
Ti	Ti^{2+}	− 1.63
Zn	Zn^{2+}	− 0.762
Cr	Cr^{2+}	− 0.6
Cr	Cr^{3+}	− 0.5
Fe	Fe^{2+}	− 0.44
Cd	Cd^{2+}	− 0.4
Co	Co^{2+}	− 0.29
Ni	Ni^{2+}	− 0.25
Sn	Sn^{2+}	− 0.136
Pb	Pb^{2+}	− 0.126
H	$\mathbf{H^{1+}}$	**0**
Sb	Sb^{3+}	+ 0.11
Cu	Cu^{2+}	+ 0.34
Hg	Hg_2^{2+}	+ 0.7986
Ag	Ag^{1+}	+ 0.7995
Pd	Pd^{2+}	+ 0.82
Hg	Hg^{2+}	+ 0.86
Pt	Pt^{2+}	+ 1.20
Au	Au^{3+}	+ 1.50

Table 11.3. Galvanic series in sea water.

Ti
70 Ni-30 Cu (monel metal)
18-8 stainless steel (passive)
Ag
C (carbon fiber reinforced plastic)
80 Ni-13 Cr-6.5 Fe (Inconel)
Ni
Cu
70-30 brass
60-40 brass
Sn
Pb
18-8 stainless steel (active)
cast iron
mild steel
Al
Zn
Mg

electrical insulation between adherends, that is provided by the adhesive layer, precludes formation of galvanic cells.

11.4 Adhesively bonded joints

Proprietary adhesives are available in the form of liquids, pastes and thin films. Most bonded joints involving composite materials are to metal structures, and it is usually surface preparation of the metal adherend that determines the performance of the joint.

The stress distribution in a bonded joint is very far from uniform. In a single lap joint, for example, almost all of the load is carried by the extreme ends of the joint. The distribution of mid-plane shear stress sketched in Figure 11.8 is, for the reasons outlined in Chapter 9, similar to that shown in Figure 9.8(c). As a consequence, the strength of an adhesive joint does not depend on the length bonded together, but it does depend on the width of the joint. Thus the pull-out loads for each of the joints sketched in Figure 11.9 are the same (compare with the stress to pull a fiber from its socket, which is independent of the fiber length, see Chapter 8, Section 8.4). This fact is just as true for a mechanical joint as it is for

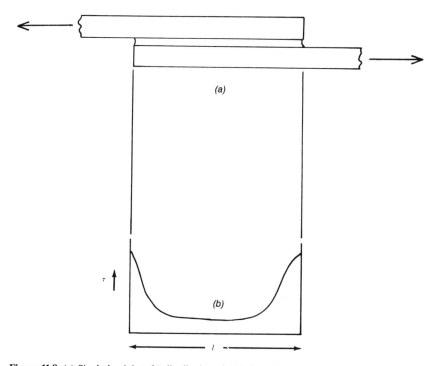

Figure 11.8. (a) Single lap joint, (b) distribution of mid-plane shear stress along the length *l* of (a).

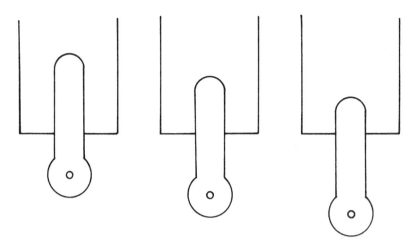

Figure 11.9. The strength of a bonded joint is independent of the depth of the bond. Each of these joints would fail at the same load.

a bonded joint; most of the load in a mechanically fastened joint is carried by the first and last bolts or rivets.

There are, in addition to the stress concentrations at the ends of a single-lap joint, other sources of failure initiation at the edges, arising from geometrical constraints peculiar to adhesive joints. In Figure 11.10, consider a tensioned single-lap joint. For the adherends to be in contact after deformation requires the

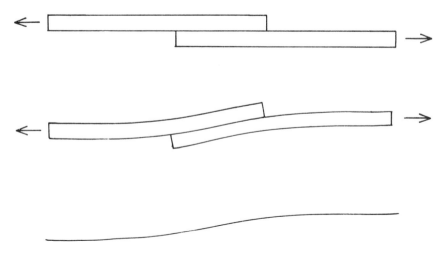

Figure 11.10. S-wise rotation in a tensioned single-lap bonded joint.

assembly to adopt S-wise bending; the lowest of the three sketches in Figure 11.10 shows this bending, with opposite curvature, at the bonded surfaces. If the overlap length is small, the joint is unable to comply with this requirement and debonding (peeling) occurs at its ends. The peel strength of proprietary adhesives are typically larger than the through-the-thickness tensile strength of fiber reinforced laminates. Hence, if one or both adherends is a laminate, delamination occurs at the second or third ply down as illustrated in Figure 11.11. The outermost plies adjacent to the adhesive are excessively tensioned and eventually fail by in-plane tensile fracture. A practical way of alleviating this effect is to taper the adherends so that the joint overall is more compliant to bending with opposite curvatures.

A somewhat similar problem arises due to swelling of the adhesive during water uptake from the atmosphere. The concentration of diffused water, and the magnitude of the swelling associated with it, is larger nearer to—than it is remote from—the edges of the adhesive layer. As a consequence, the adhesive thickens around its edges, see Figure 11.12(b). When the water content, and hence the swelling, becomes saturated at the edges, a shoulder, between uniformly thick saturated adhesive and adhesive less than saturated with water, moves inward towards the center of the joint, Figure 11.12(c). To maintain contact with the outer region of uniformly swollen adhesive, the adherends would each need to bend with curvature opposite to that inside the shoulder. Failure to adopt such S-wise bending manifests itself as an edge crack.

Thermal expansion mismatch between adherends and adhesive is another source of internal stress. The cure temperature for most adhesives is well above the operating temperature of the joint, and cooling from the adhesive cure temperature to ambient temperature, leads to distributions of stress which are subse-

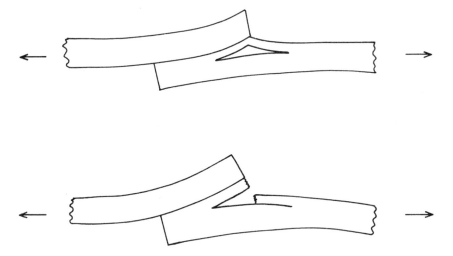

Figure 11.11. Illustrating localized delamination beneath the ends of a lap joint.

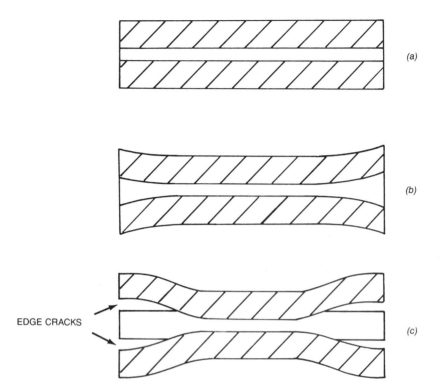

Figure 11.12. Formation of edge cracks due to inhomogeneous swelling of a butt joint.

quently relieved if the joint is used at temperatures above ambient, but accentuated if it is used at low temperatures.

Despite the S-wise bending complication, the strength of the single lap joint measured by pulling parallel to the plane of the joint is used to evaluate the performance of adhesives and adherend surface preparations. This measurement is erroneously referred to as the lap shear strength and, for most commercial adhesives, is of the order of 20 MN m^{-2} at room temperature.

11.5 Physics of adhesion

Strong chemical bonds of the kind responsible for cohesion in solids, that is, ionic, covalent and metallic bonds, are very short-ranged and might form across man-made interfaces only if interfacial gaps—between fiber and matrix in the case of composites, or between adhesive and adherends in the case of adhesive joints—can be reduced to the magnitude of interatomic distances typical for the solid state. Even then, it is by no means certain that the large changes in electronic structure necessary to form a chemical bond will be realized. After all, the re-formation of bonds between fracture surfaces brought together immediately after fracture has yet to be demonstrated for solids other than crystals which cleave between layers bonded by the relatively weaker but longer-ranged van der Waals forces. At the other extreme, the physical attachment of single gas molecules (argon for example) to clean surfaces of solids (freshly cleaved single crystal graphite for example) is both well established and well understood.

Physical interaction of gases with solids arises from the same kinds of forces that give rise to condensation of gases to form bulk liquids and solids. Consequently any gas will form an adsorbed film on the surface of any solid if the conditions of pressure and temperature are optimized. If it is assumed that the interaction energy of a gas molecule with the surface of the solid is given by the sum of the interactions of the molecule with the atoms of the solid acting as individual centers, gas-solid potential relationships can be computed by straightforward summation. For example, suppose the pair-wise gas-solid interaction has both attractive and repulsive components (see Chapter 1, Section 1.1). The interacting molecules attract each other when they are far apart and repel each other when they come close together. The force of interaction F between two spherical non-polar molecules is a function of the distance r between them. For most purposes it is more convenient to use the potential energy of interaction $\phi(r)$ rather than the force of interaction $F(r)$. The two functions are differentially related:

$$F(r) = -\frac{d\phi}{dr} \tag{11.1}$$

For non-polar molecules a commonly used intermolecular potential energy function is the Lennard-Jones[54] potential (6–12 function):

$$\phi(r) = 4\epsilon \left\{ \left(\frac{\sigma}{r}\right)^{12} - \left(\frac{\sigma}{r}\right)^{6} \right\} \tag{11.2}$$

σ and ϵ, which have dimensions of length and energy respectively, are constants characteristic of the interacting molecules. At large separations ($r > > \sigma$), the inverse sixth-power attraction component is dominant and the molecules are attracted to one another with a force proportional to the inverse seventh power of the separation. At small separations ($r < < \sigma$) the inverse twelfth-power repulsion component is dominant, and the molecules are repelled from one another with a force proportional to the inverse thirteenth power of the separation. Figure 11.13 shows the potential energy of pairs of inert gas atoms as a function of their distance apart in Å.

The overall gas-solid energy is

$$\Phi(r) = \sum_{i} \phi(r) \tag{11.3}$$

[54]J. E. Lennard-Jones. "Cohesion," *Proc. Phys. Soc.*, 43:461–482 (1931).

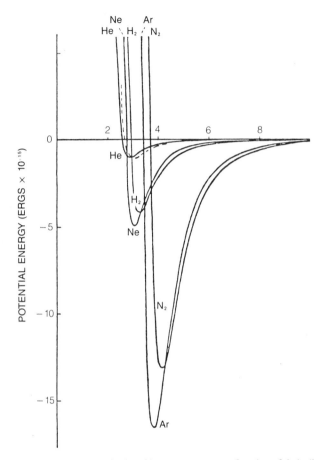

Figure 11.13. The potential energy of pairs of inert gas atoms as a function of their distance apart in Ångstroms (after Lennard-Jones[54]).

If it is further assumed that the individual centers of interaction in the solid can be smeared out to give a continuous distribution, Equation (11.2) can be replaced by an integral, the evaluation of which leads to the equation

$$\Phi(r) = \frac{2}{3} \pi n \sigma^3 \epsilon \left\{ \frac{2}{15} \left(\frac{\sigma}{z} \right)^9 - \left(\frac{\sigma}{z} \right)^3 \right\} \tag{11.4}$$

where n is the number of interacting centers in unit volume of the solid and z is the distance between the gas molecule and the solid surface.

In Chapter 8, Section 8.3, it was pointed out that, in crystalline solids, the energies of grain boundaries are of the order of 1 J m^{-2}, compared with 10^{-2} J m^{-2} for orientation twins and 10^{-3} J m^{-2} for translation twins (stacking faults). These numbers demonstrate that the smaller the departure from perfection across an internal surface, the smaller its energy. In the context of adhesive bonding to a solid, small interfacial energy likewise means near perfect chemical bonding across the interface; the solid adhesive is said to have wetted the adherend.

The term wetting is usually reserved for the interaction of liquids with solid surfaces. The idea of wetting was first conceived by T. Young,[55] to whom is attributed the well known equation defining the angle of contact θ for the liquid/vapor interface with a solid.

$$\cos \theta = \frac{\gamma_{SV} - \gamma_{SL}}{\gamma_{LV}} \qquad (11.5)$$

where γ_{SV}, γ_{SL} and γ_{LV} are the specific surface energies of the solid/vapor, solid/liquid and liquid/vapor interfaces respectively. Equation (11.5) is Young's formula.

Three classes of wetting behavior can be distinguished. The general situation where θ lies between 0 and 180°, as in Figure 11.14, is called partial wetting, or simply wetting. From Equation (11.5) it is evident that the interfacial tensions must in this case satisfy the inequality $\gamma_{SV} - \gamma_{SL} < \gamma_{LV}$. As θ tends to zero, that is as $(\gamma_{SV} - \gamma_{SL})$ approaches γ_{LV}, the liquid drop spreads over or completely wets the solid, forming in the limit a uniform thin liquid film, Figure 11.15. In the opposite limit as θ approaches 180°, that is as $(\gamma_{SL} - \gamma_{SV})$ approaches γ_{LV}, the drop ideally forms a perfect sphere tangent to the solid surface, Figure 11.16. This latter case corresponds to non-wetting. Alternatively, we could say that the vapor completely wets the solid.

If, after it has wetted the solid, the liquid itself is caused to become a solid, for example by freezing if it is water or by curing if it is an adhesive, then considerable mechanical force is required to fracture it or the interfacial bonding between it and the wetted solid. The high degree of interfacial bonding characterized by the high degree of wetting in the liquid state is carried over into the solid state.

In practice, wetting studies, for example, measurement of the angle of contact θ on photographic enlargements of sessile drops of resin adhesive, are not by themselves sufficient to account for differences in the mechanical properties of adhesive joints. Young's equation accurately describes the state of affairs if the structure at the line of contact between the three phases (adhesive, adherend, air) remains undisturbed during spreading. A real surface contains irregularities on the atomic scale, including surface steps and variations in cleanliness, and the spreading line of contact is temporarily arrested by some of these irregularities. On the other hand, if molecules evaporate from the spreading adhesive, these

[55]T. Young. "An Essay on the Cohesion of Fluids," *Phil. Trans. Roy. Soc.*, 95:65–87 (1805).

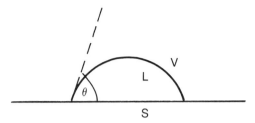

Figure 11.14.

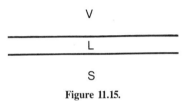

Figure 11.15.

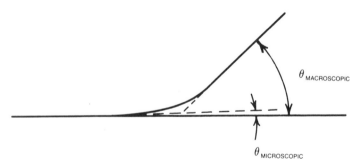

Figure 11.16.

Figure 11.17. Near-field and far-field angles of contact.

molecules move ahead of the line of contact and so pre-wet the surface. It should also be remembered that the energies in Young's equation are far-field energies. Along the line of contact between the three phases there may well be some near-field enhanced wetting, see Figure 11.17. There also exist other anomalous behaviors. Thus, a block of plasma surface treated paraffin wax is wetted by liquid resin (measured by contact angle) but will not bond to the resin when cured.

The bonding of a resin adhesive to a resin matrix composite material usually involves a measure of reptation. Reptation is the name given by P. G. de Gennes[56] to the migration of long chain molecules through polymers, and is known to occur at temperatures near and above the glass transition temperature.

Metal surfaces are usually coated with the metal oxide, which is put to good use when preparing metal surface for adhesive bonding. In the case of aluminum, for example, phosphoric acid anodizing treatments have been developed which cause columns of strongly attached hydrated aluminum oxide to grow on the metal surface. During manufacture of the joint, the resin adhesive flows into the space between the columns and makes for secure mechanical anchorage when cured.

11.6 Advanced designs for joints

Many of the fundamental problems described in Section 11.3 can be circumvented by judicial design of joints. In particular, encapsulation is widely practiced in order to constrain peeling from occurring. Two designs for joints which illustrate this philosophy are shown in Figures 11.18a and 11.18b.

Figure 11.18a is a variation of the use of internal wedges in mortise and tenon joints in carpentry. Figure 11.18b is the tapered double lap shear joint adopted to bond composite drive shafts to continuously variable (CV) joints in automobiles (see Chapter 1, Section 1.5, question #4). The problem here is that a conventional metal spigot would retract from the shaft during operation of the vehicle in sub-zero temperatures. This retraction is constrained not to occur by encapsulating the shaft inside an annular cavity in the metal.

Experience has shown that a combination of bonding and bolting produces no stronger a joint than a well-designed bonded joint. The problem with bonded-bolted joints is that, being stiffer then most bolts, the adhesive supports virtually all of the load until it fails, at which point all of the load is suddenly transferred to the bolt. If manufactured from a metal, the bolt is likely to be strain-rate sensitive in which case sudden application of load is likely to cause it to fracture in a brittle manner. The bonded/bolted combination is, however, useful for repair and for preventing the spread of damage. An example of an advanced bonded/bolted joint design is reproduced in Figure 11.19.

---

[56]P. G. de Gennes. "Reptation of a Polymer Chain in the Presence of Fixed Obstacles," *J. Chem. Phys.*, 55:572–579 (1971).

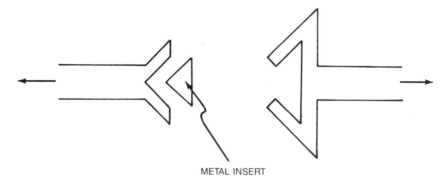

METAL INSERT

Figure 11.18a. Illustrating the use of metal inserts in joints.

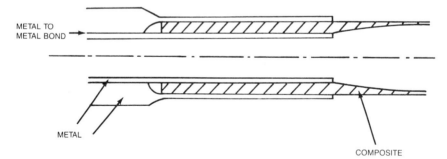

METAL TO
METAL BOND

METAL

COMPOSITE

Figure 11.18b. Joint between a composite drive shaft and a metal differential.

Figure 11.19. Stepped lap bonded/bolted joint.

11.7 Welding

In principle, it is possible to join, by welding, metal matrix composites and also thermoplastic matrix composites. One division of metallurgical welding processes is into fusion welding, for which the surfaces to be joined are melted with or without the presence of a filler metal, and pressure welding, for which heat may or may not be applied. When heat is applied during pressure welding, it may be applied from an external source, or it may be generated internally by friction between the workpieces. In either case, the melting point may or may not be reached.

The microstructure and properties of the fused zone in a weld between metal matrix components depend on metallurgical factors such as the compositions and original microstructures of the matrix and filler metals, and the heating and cooling rates. In the case of welds formed with a filler metal, the fused zone will not be reinforced by fibers and this introduces serious localized property variations.

Resistance welding is a viable technique. However, it requires the surfaces to be welded to be counter rotated in pressure contact with each other, which is possible only for butt welds. Brazing and adhesion bonding are more attractive methods for joining metal matrix compounds.

Pressure welding of thermoplastics, in the form of both composites and film adhesives, is an established joining technique.

11.8 Viscoelasticity and non-linear behavior

The resins used as adhesives are not elastic solids; that is, the change in shape imposed on an adhesive layer by tractions applied at the adherend interfaces is not wholly reversible. There is some permanent deformation arising from viscoelastic flow, the magnitude of which is a function of time, and gives rise to potential long-term problems. This behavior is typical of polymers and is somewhere between that of a viscous fluid and that of an elastic solid.

The stress in a Newtonian fluid is linearly proportional to the strain rate

$$\sigma = \eta \, \dot{\epsilon} \tag{11.6}$$

and the constant of proportionality is known as the viscosity. The stress in an elastic solid is linearly proportional to the strain

$$\sigma = c \, \epsilon \tag{11.7}$$

and the constant of proportionality is known as the stiffness. The linear combination of these equations

$$\sigma = c\,\epsilon + \eta\,\dot{\epsilon} \tag{11.8}$$

is the simplest equation describing linear visco-elastic behavior. It describes the mechanical response of a spring and dashpot in series, the so-called Voight or Kelvin element.

Symmetry between tension and compression can be expected only for linear effects. At small strains, the relationship between stress and strain is linear for most materials and is the same relationship if measured in compression as it is when measured in tension. However, the strain within a resin adhesive can be large (as it is, for that matter, in resin in the vicinity of a hole – see Chapter 4) and true symmetry between the response to compression and tension is not expected. This is true of large strain deformations in most polymers, the implications of which are described, in the context of composites for submarines, in Chapter 1, Section 1.4. Non-linear viscoelastic behavior and, in particular, its effect on fracture phenomena is a slowly emerging discipline – the fundamental principles of which are not sufficiently developed for inclusion in this book.

11.9 Repair of composite materials

Fiber reinforced airframe materials that have suffered damage in service are likely to be contaminated by a film of hydraulic fluid which, if the matrix phase of the composite is a polymer, almost certainly cannot be completely removed by cleaning. For this reason, alone, repairs are never as reliable as replacements.

For two further reasons, laminate patches are likely to be dimensionally unstable with respect to the rest of a resin matrix laminate. First, the resin in the original laminate will have post-cured and shrunk in service, and second, water uptake and the associated resin swelling will have approached saturation. The fact that the original laminate contains free water may pose yet another problem if securing the repair entails curing at a temperature above the boiling point of water. For this reason, resins with low cure temperature, but high glass transition temperature, are preferred for repair work.

Many repairs are repairs to adhesive joints. For military aircraft flying in the 1980s, the commonest airframe problem has been debonding of thin laminates from aluminum honeycomb. In most instances, corrosion of the aluminum initiates the debonding and, although removal of the corrosion products by abrasion poses no particular problem, there is usually no provision in the field for surface preparation by anodizing of the metal so exposed. By way of substitute, it is necessary to resort to electrolytic dip and, preferably, water-based primers.

Ideally, to effect repair to a damaged fiber reinforced laminate, it is necessary to provide for equalization of strains in the repaired and undamaged material. In practice this is difficult to do. Minor damage, including erosion, is usually made good using paste adhesives. Delamination, if localized, is often tackled by injec-

IMPACT

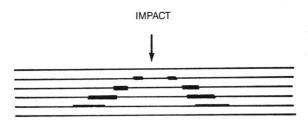

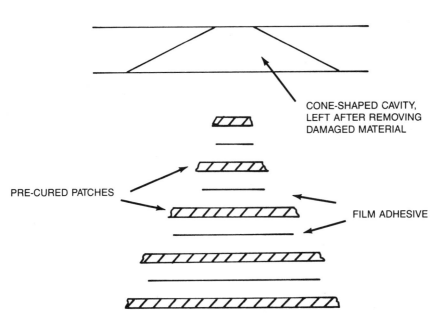

Figure 11.20. Application of adhesively bonded pre-cured laminate patches to replace the cone-shaped regions of damage produced by impact.

tion bonding using a syringe. Larger-scale damage necessitates replacement of the damage zone by bonded or bolted pre-cured patches, see Figures 11.20 and 11.21. Note, however, that if the damage is located in a curved panel, it is not possible to use a pre-cured flat patch. In all cases, the repair is mechanically isotropic or quasi-isotropic. In the case of bonded patches, quasi-isotropic laminates are commonly used, examples of the lay-up of which are illustrated in Figure 11.22. The patches shown in Figure 11.22 are unsymmetric laminates and are therefore prone to peeling on account of warping (see Chapter 7, Section 7.2). Note also that, if the laminate under repair has large Poisson's ratios as is the case for [0/±45] laminates (see Figure 6.10) then its through-the-thickness contraction will be opposed by a quasi-isotropic patch which will tend to lift off. In the case of repairs to light metal alloys, it is possible to match the stiffness of the metal by selecting a hybrid carbon, fiber/glass, fiber quasi-isotropic patch. The presence of holes lowers the effective modulus of a solid (see Chapter 1, Section 1.5, question #1). So too does the presence of cracks (see Chapter 12, Section 12.7, question #1). Repair of a hole or crack with a low modulus patch does little to make amends for the loss of modulus. On the other hand, effecting a repair with a stiff patch is to be avoided because good bonding and therefore identical strains at the bond would incur very much higher stresses in the elastically stronger patch.

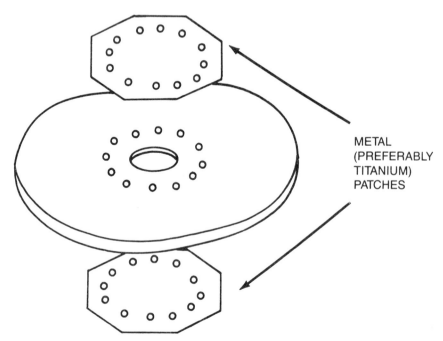

METAL
(PREFERABLY
TITANIUM)
PATCHES

Figure 11.21. Metal patches to be riveted into position.

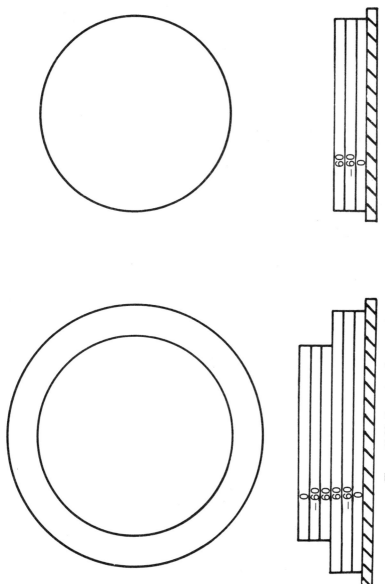

Figure 11.22. Examples of pre-cured quasi-isotropic patches (after K. T. Kedward).

11.10 Mechanical tests

When a crack propagates through an adhesive, the strain energy released comes mostly from the adherends plus, in the case of laboratory tests, the testing equipment; being a thin film, the adhesive layer has too small a volume in which to store a large amount of elastic strain energy.* The strain energy release rate ($\mathscr{G}$) or the stress intensity factor ($\mathscr{K}$) measured on an adhesive joint does not, therefore, convey very much information fundamental to the adhesive. Nevertheless, fracture toughness tests have been devised for adhesive joints: measurements for $\mathscr{G}$ are reported in the range of 100 J m^{-2} for joints made with unmodified epoxy adhesives, to 3,500 J m^{-2} for rubber modified epoxies,† and 3,200 J m^{-2} for thermoplastic polyimide, rising to 6,600 J m^{-2} for semi-crystalline polyimides. $\mathscr{K}_{Ic}$ measurements for joints manufactured with toughened epoxy adhesives are typically 1.75 MN m$^{-3/2}$. The plastic zone size predicted by Equation (8.37) for an epoxy resin of this toughness exceeds the bond line thickness, and further demonstrates that fracture mechanics tests have little to say about the properties of the adhesive itself.

One mechanical property of adhesives, which can be measured by tests that have somewhat more sound physical basis, is the peel strength. In considering a criterion for tearing of a thin sheet of cut rubber, the so-called trouser test, see Figure 11.23, R. S. Rivlin and A. G. Thomas[57] argue that consumption of the elastic strain energy released is determined mainly by the state of deformation in the neighborhood of the crotch, and that this state of deformation is determined by the shape of the cut at the crotch – not by the shape of the test-piece nor by the manner in which the tearing forces are applied. Hence, for a sheet with undeformed thickness t containing a cut of length c, extension of the cut by an increment dc requires that the work done be $\gamma t \cdot dc$, where γ is an energy per unit thickness characteristic of the material. During this incremental growth of the cut, the external forces do not move, so they do no work and the change in overall energy of the sheet is

$$- \left(\frac{\partial W}{\partial c} \right)_l = \gamma t \tag{11.9}$$

The subscript l indicates that the differentiation is carried out under constant displacement, that is under conditions when the external forces do not move. In this respect criterion (11.5) is similar to Griffith's criterion for crack extension in a

---

*This observation does not, of course, conflict with the fact that it is local forces which execute the work of fracture.

†Rubber particle toughened epoxy resins are usually manufactured by phase separation of a rubber, usually a functional butadiene-nitrile polymer. The reactive liquid rubber, initially soluble in the resin, separates out as the epoxy cures. In more recent formulations, phase separation of the rubber is caused to go to completion in the uncured epoxy resin.

[57]R. S. Rivlin and A. G. Thomas. "Rupture of Rubber. I–Characteristic Energy for Tearing," *J. Polymer Sci.*, 10:291–318 (1953).

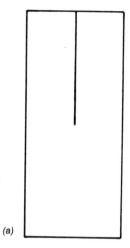

Figure 11.23. (a) Undeformed trouser test specimen.

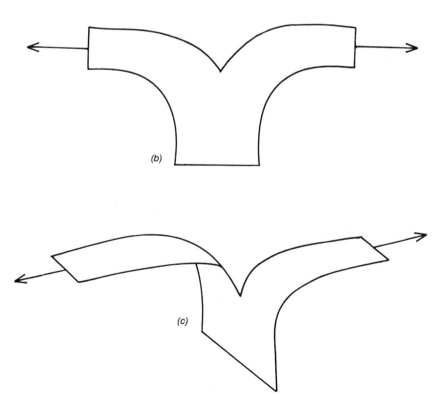

Figure 11.23. (b) and (c) Two different modes of applying tearing forces in the trouser test.

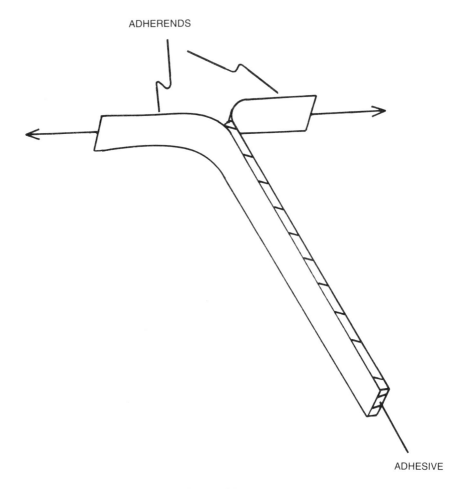

ADHERENDS

ADHESIVE

Figure 11.24. T-peel test.

plate tensioned between fixed grips, see Equation (8.13). The main difference is that γ is no longer to be interpreted as the specific fracture surface energy. Rivlin and Thomas go on to show that the force at which the trouser test-piece tears is

$$F = \frac{1}{2}\gamma t \tag{11.10}$$

The so-called T-peel test for adhesives, see Figure 11.24, similarly measures the force per unit width of bond line required to tear apart thin flexible adherends. Adhesives have notoriously small resistance to peel forces. Typical values are 500 N m^{-1} for untoughened epoxy adhesives rising to 800 N m^{-1} for rubber modified formulations.

The main criticism of the T-peel test is that the work done includes the work required to keep bending the adherends so that they remain in line with each other. Nevertheless, it does assess the peel strength of the adhesive and is therefore useful for screening purposes.

Other mechanical property measurements routinely carried out on adhesive joints include tensile modulus and tensile strength of butt joints, and strength, measured parallel to the plane of the adhesive, of single lap and double lap joints.

11.11 Worked example

1. The importance of cleaning surfaces to be bonded and, for that matter, the efficacy of application of primers by wiping, raises the question whether it is better to use a large number of small rinses (wipes) or a small number of large rinses (wipes). The same question faces Public Health Authorities when formulating guidelines for drink manufacturers who use returnable bottles. When solvent enters a bottle, dirt dissolves, but, when the bottle is emptied, a small residual volume R of solution remains, see Figure 11.25.

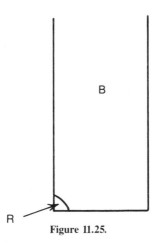

Figure 11.25.

How do you get the bottle as clean as possible, using all the solvent? Intuitively, a larger number of small rinses is expected to produce a cleaner bottle than is a small number of large rinses. Taking, as example, a single rinse of a bottle filled with cleaner versus two rinses with it half filled, if 1% of solution is always left behind when the bottle is emptied, then a dirt concentration of 1% remains after the single rinse compared with 2% of 2%, i.e., 0.04% after two rinses. Show that the volume fraction of dirt remaining is an inverse exponential function of the ratio of the volumes of solvent/residue (University of Bristol, 1985).

To generalize this argument, we shall assume that the volume W of clean solvent is very much larger than B, the volume of the bottle, which is very much larger than the small volume D of dirt initially in the bottle ($W >> B >> D$). Assume that the dirt is infinitely soluble and that several rinses are needed.

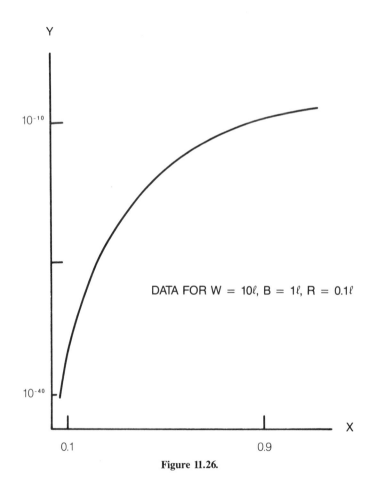

Figure 11.26.

Let $W/B = n$, and $B/R = m$

$$n \text{ and } m \text{ are both} >> 1$$

Consider an intermediate rinse for which the initial dirt content is D'.

Introduce volume xB of solvent into the bottle and let $x << 1$, i.e., partially fill the bottle. The bottle now contains volume $xB + R$ of solution. Empty the bottle. The fraction of volume D' of dirt remaining equals the volume fraction of solution left behind: $R/(xB + R)$.

Repeat this rinsing procedure with x constant. The total possible number of such rinses is $W/xB = n/x$.

When all of the solvent has been used, the fraction of dirt that remains is

$$Y = \frac{D_{final}}{D} = \left[\frac{R}{xB + R} \right]^{W/xB}$$

This function is sketched in Figure 11.26. Physically, it means that, for a given total volume of solvent used in the cleaning process, when all the solvent has been used, less dirt remains if the largest possible number of smallest volume rinses is employed.

Using a binomial series to expand the above equation, we get

$$Y = (1 + mx)^{-n/x}$$

We want to minimize Y

$$\ln Y = -\frac{n}{x} \ln(mx + 1)$$

Therefore

$$\frac{d \ln Y}{dx} = \frac{n}{x^2} \ln(mx + 1) - \frac{n}{x} \cdot \frac{m}{(mx + 1)}$$

$$= \frac{n}{x^2} \left[\ln(mx + 1) - \frac{mx}{(mx + 1)} \right]$$

What are the extremal values?

Note that the term inside the square brackets [ ] $= 0$ for $x = 0$ but that we still have $1/x^2$ outside the brackets.

It is evident that a detailed check is needed. Try expansion of $\ln Y$ for small x, that is

$$\ln(1 + \alpha) = \alpha + 0(\alpha^2) \ldots$$

$$\ln Y \cong -\frac{n}{x} [\ln x + O(x^2) \ldots)$$

$$= -nm + O(x)$$

so it looks as though there is an extremum as $x \to 0$.

The two term in $d(\ln Y)/dx$ are sketched in Figure 11.27. So, for $mx > 0$ it is concluded that $d(\ln Y)/dx > 0$ and hence $mx \to 0$ gives the minimum Y.

In the limit, and putting $Y = 1/x$, the fraction of dirt remaining is

$$\frac{D_{final}}{D} = \left(1 + \frac{m}{Y} \right)^{-Yn}$$

So, with $Y_n = \beta$ and thus $\beta \to \infty$

$$\frac{D_{final}}{D} = \lim_{\beta \to \infty} \left(1 + \frac{nm}{\beta} \right)^{-\beta}$$

$$= e^{-nm} = e^{-W/R}$$

This function for infinite solute (dirt) solubility, is shown in Figure 11.28. It is also approximately true for a large but finite solute solubility.

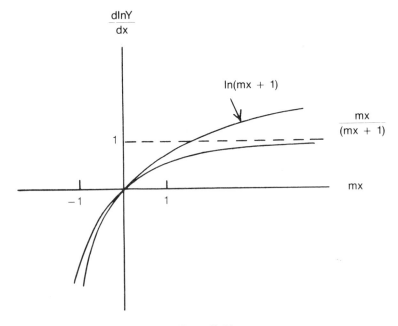

Figure 11.27.

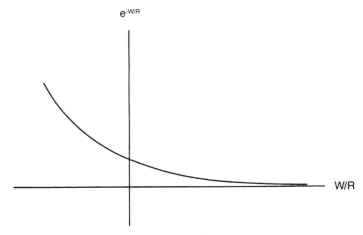

Figure 11.28.

Physically, this merely confirms the obvious; namely, that the larger the total volume of solvent, the smaller the volume fraction of dirt left behind.

11.12 Further examples

1. (library project) Refer to T. Young's[55] (1805) publication and derive from first principles the formula (11.5) now known as Young's formula.
2. (library project) An important step towards the theoretical understanding of wetting phenomena was made by J. W. Cahn, "Critical Point Wetting," *J. Chem. Phys.*, 66:3667–3672 (1977), when he drew attention to the role of phase transitions. On his desk, John Cahn has a glass phial containing two immiscible liquids, the lower one of which has risen up the wall of the phial and visibly prevents the other liquid from making contact with the glass. Referring to Figure 11.29, total displacement of fluid B (representing the air in manufac-

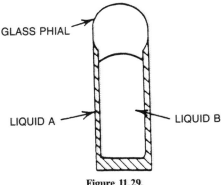

Figure 11.29.

ture of an adhesive joint) by fluid A (representing the adhesive in a joint) can be explained by postulating a wetting transition at some temperature below the critical temperature at which the fluids completely mix. By considering the thermodynamics of solutions and, in particular, the phenomenon of phase separation, account for Cahn's desktop demonstration.

12

Non-destructive evaluation

*"...cracks of a critical depth...may not always be
detected by standard ultrasonic techniques..." A. H.
Cottrell*[58]

The *raisons d'etre* for non-destructive evaluation (NDE) are (1) to meet a specification, and (2) to determine whether or not a structure is fit for use. In Chapter 8, Table 8.1 the concept of the critical Griffith crack length was introduced. This is the critical crack length which, for a given stress intensity factor, cannot be exceeded without failure. If techniques are available to detect cracks and establish whether they are shorter than this size, then, in principle, (1) and (2) can be accommodated.

The detection of cracks is a remote sensing problem; energy is directed into the structure, allowed to sample the crack, and if suitably disturbed by the crack, the disturbance may be analyzed to infer the location, size and orientation of the crack. Most forms of energy can and have been tried for fiber reinforced composites. The most practical techniques are visual, penetrants, ultrasonics, radiography and thermography.

12.1 Visual inspection

One should never ignore the great deal of information which can be gleaned by casting an eye over a fiber composite structure. Thus, the texture imparted to the surface by the underlying fibers [see Figure 5.13(a) for example] reveals the near surface fiber lay-up, of which close inspection can give an indication of uniformity and parallelism of these fibers. Delaminations which surface at the edge of a laminate can also be detected by visual examination. Other worthwhile visual observations include assessment of sources of mechanical weakness such as holes and cut outs, abrupt changes in cross-section, poor design of joints, and misalignment of joined components. For moving components fabricated from composites (such as helicopter rotor blades, turbine compressor buckets and water impellers)

---

[58]A. H. Cottrell. Addressing the February 20, 1980 meeting of the UK House of Commons, Select Committee on Energy.

it is useful to remember that loss of strength is accompanied by reduction in stiffness; so, simple periodic checks on blade deflection can be used to monitor safe residual life. Careful visual examination also assists with identification of regions to be inspected by other inspection methods.

Visual inspection of optically transparent composites can reveal internal defects. For example, the reduction in transmission of light by delaminations can be quantified using a solid state camera, and stored in a computer for construction of digitized images after subsequent propagation of the delaminated regions.

Optical holography, including holographic interferometry, is another branch of visual inspection of surface topography that can be applied to fiber reinforced materials.

12.2 Liquid penetrants

This method for revealing the presence of surface cracks exploits the rise of a liquid between two plates close together, see Figure 12.1. To estimate an order of magnitude for the height to which the liquid rises, we will assume (1) that the angle of contact is zero, that is that the liquid wets the crack faces, (2) that the crack faces are parallel and vertical, and (3) that the meniscus is semi-circular.

Consider a column of liquid of unit thickness perpendicular to the plane of the sketch in Figure 12.1. The downward force on it, that is its weight, is equal to $2T$, the sum of the two upward forces (one at the interface with each crack face) due to surface tension. The volume of the column of unit thickness is

$$xy + \left(\frac{x^2}{2} - \frac{\pi x^2}{8} \right) \tag{12.1}$$

Hence

$$\left\{ xy + \left(\frac{x^2}{2} - \frac{\pi x^2}{8} \right) \right\} \varrho g = 2T \tag{12.2}$$

where ϱ is the density of the liquid and g is the acceleration due to gravity, and so the height to which the liquid penetrates is

$$y = \frac{2T}{x \varrho g} - \left(\frac{1}{2} - \frac{\pi}{8} \right) x \tag{12.3}$$

The commonly used penetrants for fiber reinforced resins, and for polymers generally, are the organic solvents. Taking carbon tetrachloride ($\varrho \cong 1630 \text{ kg m}^{-3}$) as example, the surface tension T in contact with resins is $\cong 7\text{N m}^{-1}$. Therefore,

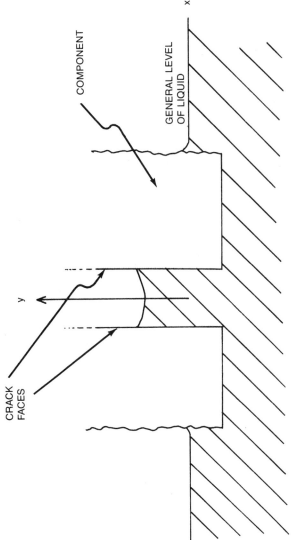

Figure 12.1. Component containing crack immersed in liquid penetrant.

for a 10μm wide crack, we have

$$y = \frac{2 \times 7 \left(\dfrac{N}{m}\right)}{10^{-5} \, (m) \times (1.62 \times 10^3) \left(\dfrac{kg}{m^3}\right) \times 9.78 \left(\dfrac{m}{s^2}\right)}$$

$$- 0.107 \times 10^{-6} \, (m)$$

$$\sim \frac{14 \left(\dfrac{N}{m}\right)}{10^{-5} \, (m) \times 15.9 \times 10^3 \left(\dfrac{N}{m^3}\right)}$$

since

$$N = \frac{m \, kg}{s^2}$$

Hence

$$y = 88 \text{ m}$$

This is for a model crack that is open to the atmosphere. In practice the compression of entrapped air considerably lowers the penetration distance but, for all practical purposes, adequate penetration is obtained.

Excess penetrant is removed by washing and, after drying the surface, the penetrant within the crack is caused to reveal itself, and hence the crack location, by dusting the surface with a fine powder known as developer. Retrieval of penetrant by the powder is also a capillary action.

12.3 Ultrasonic NDE

Elastic waves are of two kinds. P waves (primary waves) which are longitudinal compressive waves, and S waves (secondary waves) which are transverse shear waves.

In an isotropic solid the velocity of P waves is

$$V_p = \sqrt{\frac{E}{\varrho}} \tag{1.2}$$

In Chapter 1 we used V_p^2 as a measure of specific modulus and saw that, for graphite fiber for example, V_p is of the order of 10^4 m s^{-1}. The human ear is sensitive to frequencies in the range 20–20,000 Hz, the so-called audio range. Elastic waves with frequencies above 20 kHz are called ultrasound waves. The frequencies employed in conventional ultrasonic NDE are those that can be conveniently generated by the piezoelectric effect in such materials as lead zirconate titanate (PZT) and are in the range 20 kHZ to 5MHz. Using the relationship

$$\lambda = \frac{V}{\nu} \tag{12.4}$$

between wavelength (λ), velocity (V) and frequency (ν) we find

$$\lambda = \frac{10^4 (\text{m/s})}{10^6 (1/\text{s})} = 10^{-2}\text{m} = 1 \text{ cm}$$

for 1 MHz ultrasound in graphite fiber.

Wavelengths of the order of 1 cm are typical for ultrasound propagation in solids. In transparent media, glass for example, it is possible to see directly the propagation of ultrasound by viewing its associated birefringence. Thus, Figure 12.2 shows 1.5 MHz compressional waves (curved horizontal bands of birefringence) travelling from top to bottom in a block of glass. The waves are at grazing incidence to the left hand edge of the block, as the consequence of which shear waves (inclined smaller wavelength bands of birefringence) are created. The direct observation of ultrasound by this method is an invaluable teaching tool. Figure 12.2, for example, demonstrates the scale of the divergence of beams of ultrasound as well as the mode conversion phenomenon at free surfaces.

For NDE purposes, ultrasound is most commonly used in a pulse-echo mode. Short bursts of a few (say 10) cycles are generated by a transducer and transmitted, often via water in order to ensure acoustic coupling, into the material to be examined. Figure 12.3 shows the experimental set-up. The wave is reflected by the defect, for example by a region of delamination in a laminate, as well as by the far surface of the component, and returns to the transducer. The scan displayed by an oscilloscope is known in ultrasonic NDE jargon as the A-scan, Figure 12.4(a). It is simply a time domain presentation of the reflected beam (usually rectified and smoothed) for one position of the transducer.

The scan constructed from a number of A-scans recorded at points along a line [y = constant in Figure 12.4(b)] is known as a B-scan. It yields a one-dimensional trace of the defect. Similarly, the scan constructed from A-scans made across a two-dimensional xy grid, Figure 12.4(c), is known as a C-scan. The C-scan produces a real image of the defect. The depth and lateral resolution of the C-scan are determined by the length of the incident pulses and width of the beam respectively. The lateral resolution of delaminated regions routinely detected with ultrasonic C-scan is of the order of 1 cm. By using broadband electronics and transducers for the one, and focussed ultrasonic probes for the other, both limits of resolution can be made to approach the wavelength of the ultrasound.

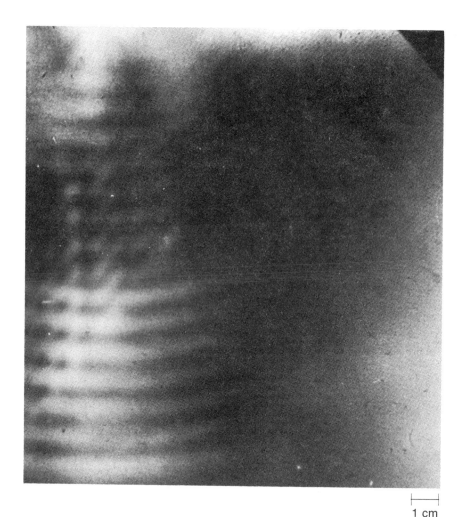

1 cm

Figure 12.2. Photoelastic visualization of ultrasonic wavefronts in glass (courtesy of T. W. Turner).

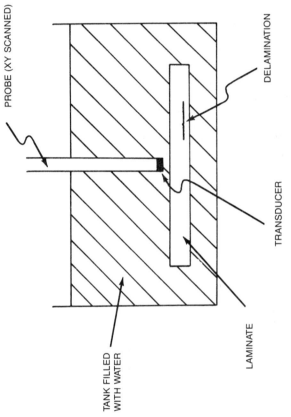

Figure 12.3. C-scan apparatus.

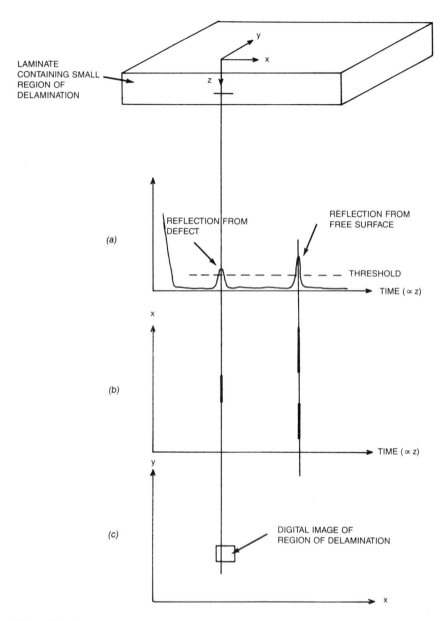

Figure 12.4. Showing the formation of (a) A-scan for which $x = y = 0$, (b) B-scan for which $y = 0$, and (c) C-scan for which $z = z_0 \pm \Delta z$.

Ultrasonic methods of non-destructive evaluation are better suited to metal matrix composites than to polymer matrix composites because ultrasound is rapidly attenuated by polymeric materials. For example, the attenuation of 7 MHz ultrasound by glass fiber reinforced epoxy resin is 3 dB cm^{-1} compared with about 10^{-3} dB cm^{-1} for metal matrix composites.

12.4 X-radiography, γ-radiography and neutron radiography

Consider a monochromatic (one wavelength) beam of X-rays, γ-rays or neutrons propagating through matter of any kind (solid, liquid or gas). Due to absorption, the intensity I is diminished by dI in distance dx. dI/I is linearly proportional to dx,

$$\frac{dI}{I} = -\mu dx \tag{12.5}$$

where the constant of proportionality μ is known as the linear absorption coefficient. The magnitude of μ depends on the wavelength and on the material. Values for some chemical elements of interest in composite materials are listed in Tables 12.1 and 12.2.

It is evident that absorption of X-rays is higher the higher the atomic number which, since the fiber and matrix material in composites are usually lightweight, means that X-radiography cannot easily resolve one from the other. The same is not true for neutron radiography; boron, for example, heavily absorbs neutrons on account of the (n, α) reaction with B^{10}. Thus boron fibers and particularly the bunching together of boron fibers can be seen in neutron radiographs.

Integrating Equation (12.6) we get

$$I = I_0\, e^{-\mu x} \tag{12.6}$$

which says that the transmitted intensity falls off exponentially with thickness of material traversed.

A useful parameter for comparing absorption by different materials is the

Table 12.1. Absorption coefficients μ (cm^{-1}) for 0.098 Å wavelength X-rays.

boron	0.35
carbon	0.33
aluminum	0.42
titanium	0.98
tungsten	56.00

Table 12.2. Absorption coefficients μ (cm^{-1})
for 1.08 Å wavelength neutrons.

boron	60.00
carbon	0.60
aluminum	0.97
titanium	0.54
tungsten	1.10

thickness at which the incident intensity is reduced by one half, that is

$$\frac{I}{I_0} = \frac{1}{2} = e^{-\mu x} = \frac{1}{e^{\mu x}} \tag{12.7}$$

Rearranging (12.7) we get

$$e^{\mu x} = 2$$

Taking logarithms,

$$\mu x = \ln (2)$$

$$= 0.69$$

So the half-thickness

$$x_{1/2} = \frac{0.69}{\mu} \tag{12.8}$$

Table 12.3 lists the half-thicknesses for several materials of interest to composites, irradiated with X-rays of three different wavelengths.

Of particular interest is the large difference in half-thickness between compos-

Table 12.3. Thickness of material $x_{1/2}$ (cm) at which X-ray intensities
are reduced to half the incident intensity.

	$\lambda = 0.1$ Å	$\lambda = 0.7$ Å	$\lambda = 2.0$ Å
air	—	410	26
polymer (cellophane)	4.3	0.4	0.05
carbon	2.1	—	—
aluminum	1.6	0.05	0.0025

ites, $\cong 3$ cm for carbon fiber reinforced resins, and air, effectively infinity for hard (short wavelength) X-rays. It is this difference which makes possible the detection of voids in composite materials. The generation of hard X-rays requires high voltage ($> 1,000$ keV) equipment. A less expensive source of high penetration radiation (γ-rays) are the radioactive isotopes, for example radium ($\lambda < 0.05$ Å). An additional advantage of isotopic γ-ray sources is that they are small and portable and can, for example, be easily inserted inside hollow tubes.

To enhance the contrast at defects which emerge at the external surface, it is sometimes worthwhile filling the defect with radio opaque material such as a barium salt. The relative absorption giving rise to the contrast is then that between the composite and a heavy metal, rather than that between the composite and air. This technique is known as penetrant enhanced radiography.

12.5 Thermography

The term "thermographic" was first invented to describe the temperature recording application of the Bourdon gage. Its meaning has changed over the years and is now synonymous with liquid crystal films sandwiched between microscope cover slips (or similar) used to detect small ($<1°$) local changes in skin temperature, and hence locate tumors in the human body. When observed by reflected light on a dark ground, many liquid crystals, particularly the smectics, exhibit beautifully colored patterns of birefringence arising from the long-range correlation in molecular orientation; the refractive index parallel to any arbitrary axis in the molecule is different from that perpendicular to it. The observed regions of rapidly changing birefringence coincide with regions of rapidly changing twist period of the order of the wavelength of light between neighboring liquid crystals. They are very sensitive to temperature changes.

Liquid crystal thermography can detect local small variations (increases and decreases) in thermal diffusivity arising from inhomogeneities in composite materials. Two examples are illustrated in Figure 12.5(a) and (b). In both cases, the laminate is illuminated by a source of heat such as an infrared lamp. In Figure 12.5(a) the reduction in thermal diffusivity at a discontinuity such as a delamination is revealed by a color change in the liquid crystal that corresponds to a cold spot. In Figure 12.5(b), the opposite is the case. Local densification in the damaged zone of a laminate has raised the thermal diffusivity and the damage is detected as a local hot spot.

The advent of ultrasensitive liquid crystal thermography has made possible the detection in composite materials of the adiabatic heating that accompanies alterations in strain. Referring to Figure 12.6, a length l_o of fiber is suddenly subjected to a tensile load W. The immediate response is adiabatic although, after a time, the fiber will reach equilibrium with its surroundings.

$$dU = Wdl + TdS \qquad (12.9)$$

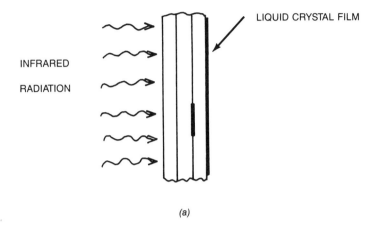

(a)

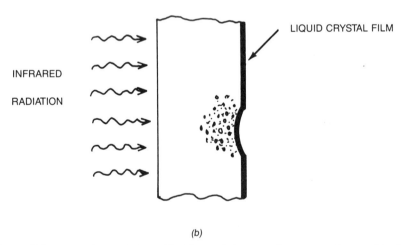

(b)

Figure 12.5. (a) and (b) Application of liquid crystal thermography to composite materials. The liquid crystal film has to be held in close proximity to the surface remote from the incident heat.

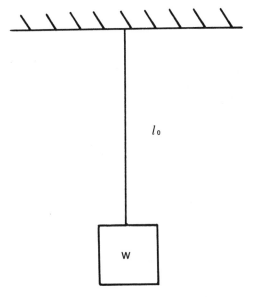

Figure 12.6. Illustrating adiabatic heating of a fiber.

We can write, for the temperature change for adiabatic reversible change,

$$\left(\frac{\partial T}{\partial W}\right)_s = -\left(\frac{\partial l}{\partial S}\right)_w$$

$$= \left(\frac{\partial l}{\partial T}\right)_w \bigg/ \left(\frac{\partial S}{\partial T}\right)_w \qquad (12.10)$$

For small loads, the numerator is a coefficient of thermal expansion and the denominator is $1/T$ multiplied by a specific heat at constant pressure.

Assuming a linear coefficient of expansion and Hooke's law, we have the following linear relationship between length, temperature and load:

$$l = l_o\{1 + \alpha(W)T + \beta(T)W\} \qquad (12.11)$$

so

$$\frac{\partial l}{\partial T} = l_o\left(\alpha + W\frac{\partial \beta}{\partial T}\right) \qquad (12.12)$$

To obtain an expression for $(\partial S/\partial T)_w$ we make use of the Maxwell reciprocal relationship

$$\left(\frac{\partial S}{\partial W}\right)_T = \left(\frac{\partial l}{\partial T}\right)_W = l_o\left(\alpha + W\frac{d\beta}{dT}\right) \tag{12.13}$$

Integrating with respect to W, we find

$$S = S_o + l_o\left[\int_o^W \alpha\, dW + \frac{1}{2}W^2\frac{d\beta}{dT}\right] \tag{12.14}$$

Hence

$$\left(\frac{\partial S}{\partial T}\right)_W = \frac{dS_o}{dT} + l_o\frac{1}{2}W^2\frac{d^2\beta}{dT^2}$$

$$= \frac{l_o a_o \varrho_o c_p}{T} + l_o\frac{1}{2}W^2\frac{d^2\beta}{dT^2} \tag{12.15}$$

where a_o is the original cross-section of the fiber and ϱ is its original density. Therefore

$$\left(\frac{dT}{dW}\right)_S = -\frac{\alpha + W\dfrac{d\beta}{dT}}{\dfrac{a_o\varrho_o c_p}{T} + \dfrac{1}{2}W^2\dfrac{d^2\beta}{dT^2}}$$

or,

$$\left(\frac{dT}{d(W/a_o)}\right)_S = -\frac{\alpha + W\dfrac{d\beta}{dT}}{\dfrac{\varrho_o c_p}{T} + \dfrac{1}{2a_o}W^2\dfrac{d^2\beta}{dT^2}} \tag{12.16}$$

$W/a_o = \sigma$ is the applied stress, α is the thermal expansion coefficient, $\varrho_o c_p$ is a volumetric specific heat and $d\beta/dT$ and $d^2\beta/dT^2$ are temperature coefficients of elastic compliance. At low loads,

$$\left(\frac{dT}{d\sigma}\right)_S = -\frac{\alpha T}{\varrho_o c_p} \tag{12.17}$$

For fiberglass, $\alpha \cong 4 \times 10^{-6}$ fractional extension per Kelvin, $\varrho_o \cong 2.2 \times 10^3$

kg m^{-3} and $c_p \cong 500$ J kg^{-1} K^{-1}. Hence,

$$\left(\frac{dT}{d\sigma}\right)_s \cong - 10^{-9} \left(\frac{K}{N\ m^{-2}}\right) \tag{12.18}$$

At 300 K, the strength of fiberglass is of the order of 3 GN m^{-2}; so, adiabatic loading up to the elastic limit would drop the temperature by about three degrees. The above analysis is due to Lord Kelvin.

12.6 Fiber optic sensors

The earliest published attempt to use fiber optic sensors to monitor simulated in-service changes in composite materials was by S. Y. Field and K. H. G. Ashbee.[59] They used fibers drawn from a lead glass as light pipes in order to follow the development of osmotic pressure filled cracks at fiber/polyester resin interfaces. The pressure due to osmosis can be several hundred bars (see Chapter 10, Section 10.4) and is sufficient to locally densify the resin and hence increase the refractive index of the resin so that it exceeds that of the fiberglass. When this phenomenon occurs, light is lost to the resin and the intensity emerging from the fiber is attenuated.

Since this early experiment, the feasibility has been demonstrated of using fiber optics on large scale structures, such as space stations, to sense degradation, temperature, pressure, acceleration, strain, acoustic emission and ionizing radiation. Since they contain their own sensors for materials evaluation, such structures have been dubbed "smart," meaning intelligent.

12.7 Worked example

1. It is proposed to use velocity of sound measurements to monitor the progress of microcracking including delamination in a large submarine structure. Investigate the effects on velocity ratio v_p/v_s, where v_p and v_s are the P-wave and S-wave velocities, respectively, of the occurrence of microcracking, and of subsequent seepage of water into the microcracks. You may ignore elastic anisotropy.

Sound, including ultrasound, is propagated by two modes. The primary wave (P) is a longitudinal or compressional wave which travels by way of elastic displacements parallel to the direction of propagation. The secondary wave (S) is a transverse or shear wave which travels by way of elastic displacements perpendicular to the direction of propagation. In an elastically isotropic medium, the P-

---

[59]S. Y. Field and K. H. G. Ashbee. "Weathering of Fiber Reinforced Plastics: Progress of Debonding Detected in Model Systems by Using Fibers as Light Pipes," *Polymer. Engrg. and Science*, 12:30–33 (1972).

wave and S-wave velocities are

$$
\left.
\begin{aligned}
v_P &= \sqrt{\frac{\lambda + 2\mu}{\varrho}} \\[2ex]
v_S &= \sqrt{\frac{\mu}{\varrho}}
\end{aligned}
\right\}
\tag{12.19}
$$

and

where λ and μ are Lamé's constants of elasticity. Thus,

$$
\left(\frac{v_P}{v_s}\right)^2 = \frac{\lambda + 2\mu}{\mu}
$$

$$
= \frac{\varkappa}{\mu} + \frac{4}{3}
$$

since

$$
\lambda = \varkappa - \frac{2\mu}{3}
$$

Substituting for $\varkappa$ and μ, we get

$$
\left(\frac{v_P}{v_s}\right)^2 = \frac{2(1 + \nu)}{3(1 - 2\nu)} + \frac{4}{3}
$$

$$
= \frac{2(1 - \nu)}{(1 - 2\nu)}
\tag{12.20}
$$

That is, the sound velocity ratio depends only on Poisson's ratio ν.

The creation of cracks decreases the Poisson's ratio and hence decreases the velocity ratio, an effect which is measurable in the case of seismic wave propagation prior to earthquakes. Referring to Figure 12.7, it is intuitively expected that the Young's modulus measured perpendicular to a crack will be less than normal, whereas that measured parallel to the crack will be unaffected. The shear modulus, however, is likely to be unaffected perpendicular to the crack but reduced parallel to it. The equations reported by J. R. Bristow[60] (with one misprint corrected) predict that, for n randomly oriented penny-shaped cracks per unit

---

[60]J. R. Bristow. "Microcracks, and the Static and Dynamic Elastic Constants of Annealed and Heavily Cold-Worked Metals," *Brit. J. Appl. Phys.*, 11:81–85 (1960).

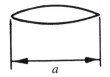

Figure 12.7.

volume, each of radius a,

$$\frac{\Delta E}{E} = \frac{16}{45} \cdot \frac{(1 - \nu^2)(10 - 3\nu)}{(2 - \nu)} a^3 n \tag{12.21}$$

$$\frac{\Delta \mu}{\mu} = \frac{32}{45} \cdot \frac{(1 - \nu)(5 - \nu)}{(2 - \nu)} a^3 n \tag{12.22}$$

$$\frac{\Delta \varkappa}{\varkappa} = \frac{16}{9} \cdot \frac{(1 - \nu^2)}{(1 - 2\nu)} a^3 n \tag{12.23}$$

Hence, if $\nu = 1/3$, $\varkappa$ is decreased at a rate which is three and a half times as fast as the rate of decrease in μ. A decrease in the ratio v_p/v_s from 1.75 to 1.5, for example, could be attributed to a crack density of $\cong (1/10)a^3$ per unit volume, where a is the average crack radius.

By virtue of drainage from nearby cracks as well as from the free surface, freshly created cracks will surely become filled with water and $\varkappa$ should eventually increase but μ should remain unchanged. That is, seepage of water into microcracks is expected to produce a recovery in velocity ratio.

The use of velocity ratio measurements to study microcracking in small structures including laboratory-scale structures is of considerable interest. However, it should be remembered that, unlike the rod-wave, the P-wave is characterized by a wavelength very much shorter than the specimen length and is not associated with a Poisson contraction.

12.8 Further examples

1. For what geometric reason is it usually not possible to detect delaminations by conventional X-radiography?
2. Why are very low accelerating voltages (albeit, with long exposure times) best suited for detection by X-radiography of voids in polymeric materials?
3. To detect near surface defects, it is usual to employ high frequency ultrasound. Why?
4. Figure 12.8(a) shows a shaft immersed in water and illuminated with a wide parallel beam of ultrasound emitted by transducer A. The shaft contains a defect at B and, as a consequence, the first four echoes occur at the relative times shown on the oscilloscope trace reproduced in Figure 12.8(b). Sketch

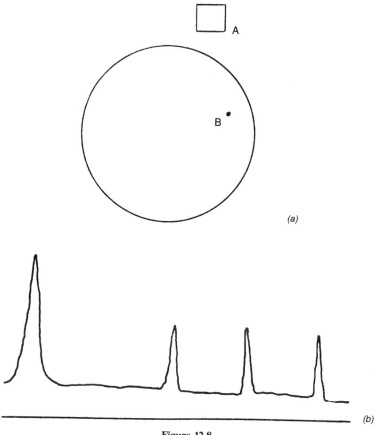

(a)

(b)

Figure 12.8.

the ray paths for these first four echoes and identify the mode propagation along each ray.

5. Using a puff of compressed air, a single transverse wave is launched along a tensioned rope. By graphical construction of the displacement versus time relationship, or otherwise, explain why successive reflections from a severely worn section reinforce one another to give a progressively increasing signal from the defective region (after H. Kwan and G. L. Burkhardt, SWRI).

6. Some metal matrix composites rapidly attenuate ultrasound, which is usually attributed to one or more the following:

  (i) Coarse matrix dendrites
  (ii) Coarse matrix grain size in the vicinity of welds
 (iii) Fiber bunching

What is the origin of the attenuation, and why and how does it depend on the size of the dendrites, grains, and fiber bundles?

7. (Library assignment) Laminates are characterised by two well defined lengths, namely the fiber diameter d and the ply thickness h. The propagation through laminates of ultrasound for which the wavelength $\lambda \sim h \gg d$ is a problem of wave propagation in periodic structures. Discuss the origin of dispersion, including the occurrence of passing bands and stopping bands. What is meant by the terms "acoustic branch" and "optical branch" of the dispersion surface? In the case of polymer matrix laminates, what is viscoelastic dispersion?

8. Because standing waves of ultrasound are formed in a tank full of oil, the velocity of light at a certain instant varies with position as $c(x, y, z) = \cos(1 + 10^{-4} \cos \{\pi x/\mu m^2\})$. Without a detailed derivation of the form of the dispersion surface sketch, giving as many numerical values as possible, the distribution of light around a small monochromatic source ($\lambda = 1\mu m$) immersed in the tank. How would the distribution of light be altered if the drive to the ultrasonic transducer was a saw-tooth instead of a sinusoid (University of Bristol Examination, 1975)?

9. Polyester resin reinforced with low volume fractions of fiberglass is used extensively as roofing panels. In order to make the fibers "invisible" the manufacturers usually try to match the refractive index of the resin with that of the fiber (~ 1.5 for E glass fiber). Two or three years after installation, these materials undergo changes known in the trade as "fiber whitening" or "fiber prominence," the principal cause of which is development of physical gaps at the fiber/resin interface. By constructing ray diagrams, account for this phenomenon and suggest reasons for its occurrence.
Very small ($< \lambda/2$) interfacial gaps accommodate frustrated total internal reflection. What is this phenomenon and how could it be exploited as a tool for NDE?

10. Optical waveguides have substantially larger diameters than do fibers used to reinforce plastics. They also have very different physical properties (thermal conductivity, specific heat capacity, coefficient of thermal expansion, electrical conductivity) from those of reinforcing fibers other than fiberglass. Discuss the implications of these differences on the possibility of using embedded optical waveguides to monitor temperature, pressure, acceleration, strain, acoustic noise, and ionization in laminates used in aircraft and spacecraft.

Suggested reading

In my university, the library bookshelves afforded to fiber reinforced composites contain many bound volumes of conference proceedings and the like, but very few genuine textbooks written for teaching purposes. Of that few, I recommend the following to my students:

R. F. S. Hearman. *Applied Anisotropic Elasticity*, Oxford University Press (1961).

B. D. Agarwal and L. J. Broutman. *Analysis and Performance of Fiber Composites*, John Wiley and Sons (1980).

A. Kelly and N. H. Macmillan. *Strong Solids 3rd edition*, Clarendon Press, Oxford (1986).

R. M. Jones. *Mechanics of Composite Materials*, McGraw-Hill, New York (1975).

D. Hull. *An Introduction to Composite Materials*, Cambridge University Press (1981).

M. R. Piggott. *Load-Bearing Composite Materials*, Pergamon Press, Oxford (1980).

J. C. Halpin. *Primer on Composite Materials*, Lancaster, PA:Technomic Publishing Co., Inc. (1984).

J. M. Whitney. *Structural Analysis of Laminated Anisotropic Plates*, Lancaster, PA: Technomic Publishing Co., Inc. (1987).

S. W. Tsai and H. T. Hahn. *Introduction to Composite Materials*, Lancaster, PA: Technomic Publishing Co., Inc. (1980).

Index